Managing Data From Knowledge Bases:
Querying and Extraction

Wei Emma Zhang • Quan Z. Sheng

Managing Data From Knowledge Bases: Querying and Extraction

 Springer

Wei Emma Zhang (iD)
Department of Computing
Macquarie University
Sydney, NSW, Australia

Quan Z. Sheng
Department of Computing
Macquarie University
Sydney, NSW, Australia

ISBN 978-3-030-06940-7   ISBN 978-3-319-94935-2   (eBook)
https://doi.org/10.1007/978-3-319-94935-2

Printed on acid-free paper

This Springer imprint is published by the registered company Springer Nature Switzerland AG
The registered company address is: Gewerbestrasse 11, 6330 Cham, Switzerland

Foreword

Knowledge bases (KBs) are the most essential components in realizing semantic computing for better human-machine interaction experiences. Knowledge bases supply facts and relationships for use in computation by machines. This can facilitate artificial intelligence (AI) tools with the ability to reason and explain. Over the years, knowledge base has been receiving much attention, both from academia and industry, as a resource for providing knowledge, an auxiliary tool for facilitating the searching on search engines, and an expert system for helping in decision making.

Knowledge available for improving computations by AI tools has grown to become quite large, which presents a number of technical challenges including efficient knowledge retrieval and automatic knowledge base construction. Among the books on the market that cover various challenges related to KBs, this book presents one of the rare attempts to present innovative solutions for the knowledge extraction and querying in knowledge bases.

These topics are under the umbrella of extracting knowledge from unstructured data for the effective construction of knowledge bases and querying knowledge bases based on a learning-based cache framework. The book overviews key findings from the authors' intensive research experience in analyzing data from different knowledge sources for knowledge base queries and knowledge base construction. The extensive references included in this book will help the interested readers find out more information on the discussed topics.

I am happy to commend the authors for their outstanding accomplishment and to inform the readers that they are looking at an authoritative piece of work in the vibrant and rapidly expanding field of knowledge extraction and discovery. This book is a valuable resource for everyone interested in the topics this book covers in depth.

Dayton, OH, USA
April 2018

Amit Sheth

Preface

Semantic Web is a paradigm that publishes and retrieves knowledge on the Web in a semantically structured way. Knowledge base (KB) is one of the most essential components in realizing the idea of Semantic Web as it provides facts and relationships that can be automatically understood, interpreted, and deduced by machines (e.g., programmatic software). Recently, knowledge base has gained momentum in providing accurate, expert, and multidisciplinary knowledge to the society. While it is well understood that knowledge base offers numerous opportunities and benefits, it also presents significant technical challenges. Among them, effective and efficient knowledge extraction and retrieval are two fundamental challenges facing the research community and industry today.

In this book, we first address the research issues and explore the principles and techniques of the challenging topics. Then we solve the raised research issues by developing a series of methodologies. More specifically, we study the query optimization and tackle the query performance prediction for knowledge retrieval. We also handle unstructured data processing and data clustering for knowledge extraction. To optimize the queries issued through interfaces against knowledge bases, we propose a cache-based optimization layer between consumers and the querying interface to facilitate the querying and solve the latency issue. The cache depends on a novel learning method that considers the querying patterns from individual's historical queries without having knowledge of the backing systems of the knowledge base. To predict the query performance for appropriate query scheduling, we examine the queries' structural and syntactical features and apply multiple widely adopted prediction models. Our feature modeling approach eschews the knowledge requirement on both the querying languages and system. To extract knowledge from unstructured Web sources, we examine two kinds of Web sources containing unstructured data: the source code from Web repositories and the posts in programming question-answering communities. We use natural language processing techniques to pre-process the source codes and obtain the natural language elements. Then we apply traditional knowledge extraction techniques to extract knowledge. For the data from programming question-answering communities, we make the attempt towards building programming knowledge base by starting with paraphrase identification problem and develop novel features to accurately identify duplicate posts. For domain-specific knowledge extraction, we propose to use clustering technique to separate knowledge into different groups. We focus on developing a new clustering algorithm that uses manifold constraint in the optimization task and achieves fast and accurate performance. For each of model and approach presented in this book, we have conducted extensive experiments to evaluate it using either public dataset or synthetic data we generated. We also discuss some open research directions at the end of this book.

Sydney, NSW, Australia   Wei Emma Zhang
April 2018   Quan Z. Sheng

Acknowledgments

I would like to express my gratitude to Prof. Michael Sheng, Prof. Mingkui Tan, and A/Prof. Kerry Taylor for their help and guidance. I owe a huge debt to my parents, my husband, and my prince for their support and sacrifice. They have been always there for me whenever I needed them. Without them, I could not be successful at any point of time.

Wei Emma Zhang

To my mum for her love.

Quan Z. Sheng

Contents

1   Introduction 1
  1.1   Overview of Knowledge Bases 2
  1.2   Overview of Knowledge Extraction in Knowledge Bases 4
     1.2.1   Extraction Techniques Overview 4
     1.2.2   Representation Models Overview 7
  1.3   Overview of Knowledge Bases Question Answering 10
     1.3.1   Question Answering in Curated KBs 10
     1.3.2   Question Answering in Open KBs 14
  1.4   Research Issues in Querying and Extracting Knowledge Bases 14
     1.4.1   An Architecture for Knowledge Base Management 15
     1.4.2   Our Contributions 16
  1.5   Book Organization 18

2   Cache Based Optimization for Querying Curated Knowledge Bases 19
  2.1   Design Overview 19
  2.2   The SPARQL Endpoint Cache Framework 23
     2.2.1   Query Distance Calculation 24
     2.2.2   Feature Modelling 28
     2.2.3   Suggesting and Prefetching Similar Queries 29
     2.2.4   Caching and Replacement 31
  2.3   Experimental Evaluations and Discussions 34
     2.3.1   Setup 34
     2.3.2   Analysis of Real-World SPARQL Queries 35
     2.3.3   Performance of Cache Replacement Algorithm 37
     2.3.4   Comparison of Feature Modelling Approaches 39
     2.3.5   Performance Comparison with the State-of-the-Art 42
     2.3.6   Discussions 43
  2.4   Related Work 44
     2.4.1   Semantic Caching 44
     2.4.2   Query Suggestion 45
  2.5   Summary 46

3   Query Performance Prediction on Knowledge Base 47
    3.1   Design Overview 47
          3.1.1   Motivation 48
          3.1.2   Challenges 49
          3.1.3   Prediction Approach Overview 50
    3.2   Preliminaries 52
          3.2.1   Multiple Regression 52
          3.2.2   Dimension Reduction 53
    3.3   Feature Modelling for Queries 53
          3.3.1   Algebra Features 53
          3.3.2   BGP Features 55
          3.3.3   Hybrid Features 55
    3.4   Predicting Query Performance 56
          3.4.1   Predictive Models 56
          3.4.2   Two-Step Prediction 57
    3.5   Experimental Evaluation and Discussion 58
          3.5.1   Setup 58
          3.5.2   Prediction Models Comparison 59
          3.5.3   Feature Modelling Comparison 60
          3.5.4   Comparison of Different Weighting Schemes in k-NN
                  Regression 62
          3.5.5   Performance of Two-Step Prediction 63
          3.5.6   Comparison to State-of-the-Art 64
    3.6   Discussions 65
    3.7   Related Work 66
          3.7.1   Query Performance Prediction via Machine Learning
                  Algorithms 66
          3.7.2   SPARQL Query Optimization 67
    3.8   Summary 67

4   An Efficient Knowledge Clustering Algorithm 69
    4.1   Overview of Clustering with Non-negative Matrix Factorization 69
    4.2   Orthogonal Non-negative Matrix Factorization Over Stiefel
          Manifold 71
          4.2.1   Notations 71
          4.2.2   Optimization on Stiefel Manifold 71
          4.2.3   Update U via NRCG 73
          4.2.4   Update V 76
          4.2.5   Convergence Analysis 78
    4.3   Experimental Evaluation 78
          4.3.1   Implementation Details 79
          4.3.2   Data Sets 79
          4.3.3   Metrics 81
          4.3.4   Results 82
    4.4   Related Works 87
    4.5   Summary 88

5   Knowledge Extraction from Unstructured Data on the Web 89
  5.1   Design Overview 89
  5.2   Source Code Topics Extraction via Topic Model and Words
      Embedding 91
     5.2.1   Data Pre-processing 92
     5.2.2   Topic Extraction 92
     5.2.3   The Coherence Measurement 95
     5.2.4   Automated Terms Selection for Topic Extraction 96
  5.3   Experimental Evaluation 97
     5.3.1   Setup 97
     5.3.2   Results 98
  5.4   Related Works 101
  5.5   Summary 102

6   Building Knowledge Bases from Unstructured Data on the Web 103
  6.1   Design Overview 103
  6.2   Prototype of Knowledge Extraction from Programming
      Question Answering Communities 105
     6.2.1   Question Extraction 106
     6.2.2   Answer and Tags Extraction 106
     6.2.3   Triple Generation 106
  6.3   Detecting Duplicate Posts in Programming QA Communities 107
     6.3.1   Pre-processing 107
     6.3.2   Feature Modelling 108
     6.3.3   Binary Classification 111
  6.4   Experimental Evaluation and Discussions 112
     6.4.1   Setup 112
     6.4.2   Results 114
     6.4.3   Discussions 119
  6.5   Related Work 120
     6.5.1   Question Retrieval from QA Communities 120
     6.5.2   Mining PCQA Websites 121
  6.6   Summary 121

7   Conclusion 123
  7.1   Summary 123
  7.2   Future Directions 125

References 127

Chapter 1
Introduction

Semantic Web is a paradigm which publishes and queries knowledge on the Web in a semantically structured way [1]. Tracing the usage of the term "Semantic Web" in the computer science field, it comes to Tim Berners-Lee's keynote speech at the 1995 International Networking Conference (INET'95), in which he outlined four points on the way to a "Semantic Web" [2]: (1) Link typing; (2) Knowledge representation content types; (3) Meta language for trait investigation and (4) Bootstrapping class structures. The terminology then was formalized and introduced to a wider audience by Tim Berners-Lee in 2001 [3]: Semantic Web is a Web of Data where not only documents and links, but the entities and relations are represented.

Knowledge base (KB) is one of the most essential components in realizing the idea of Semantic Web as it provides facts and relationships that can be automatically understood by machines (e.g., programmatic software). Knowledge bases are used for collection, organization, retrieval and sharing of the knowledge as well as being an artificial intelligence tool for giving explanation and reasoning support [4]. Over the past few years, knowledge base has been receiving much attention as a resource for providing knowledge, an auxiliary tool for facilitating the searching on search engines, an expert system for helping decision making in various domains. The advances in information extraction have led to the construction of web-scale structured and linked semantic knowledge bases (e.g., DBPedia [5], YAGO [6] and WikiData [7]). These knowledge bases cover knowledge of entities in a great range from people to organizations, from cities to countries and from animals to species etc. Besides the general knowledge bases, domain-specific knowledge bases such as GeneOntology [8] and WordNet [9] provide expert knowledge to a small group of people and make the knowledge bases widely applicable: people could find both general and domain-specific knowledge by question against the knowledge bases rather than searching and browsing from Web documents. The expertise and generality of knowledge bases motivates numerous applications in the industry. Google' Knowledge Graph is a large knowledge base that contains over 70 billion

© Springer International Publishing AG, part of Springer Nature 2018
W. E. Zhang, Q. Z. Sheng, *Managing Data From Knowledge Bases:
Querying and Extraction*, https://doi.org/10.1007/978-3-319-94935-2_1

facts about people, places and organizations until October 2016 [10]. The scalable knowledge base is the backbone of many Google applications and helps Google to deliver more semantic search results. Similar to Google, Microsoft's Satori is a knowledge base to enhance Bing's semantic searching [11]. Facebook is also building a knowledge base, called "Entity Graph" from their users data, aiming at boosting the social network searching [12]. Very recently, LinkedIn reveals its knowledge base build upon LinkedIn's entities such as member, companies, skills, certificates [13].

Given the accelerated adoption of knowledge bases in the industry, the accurate and efficient knowledge retrieval is still considered as one of the key challenges in the adoption of knowledge bases. Indeed, the knowledge consumers rely on the knowledge obtained from knowledge bases to make their decisions. Querying on knowledge bases that use entity relational model as data model and ontologies to represent knowledge is essentially querying with the structured query language (e.g., SPARQL) supported by the underlying databases. Many knowledge bases support protocol services (e.g., SPARQL endpoint) to enable the querying from public available interfaces. However, querying from these protocol services are not efficient due to the reasons such as network instability and latency. Although most research efforts have been put on querying localized knowledge bases (i.e., the querying is performed directly on the underlying databases), several researchers have recognized the importance of querying public available knowledge bases through protocol services and proposed solutions to tackle the efficiency issue [14–17].

The efficient and accurate knowledge retrieval from knowledge bases also relies on the quality of extracted knowledge. The heterogeneity of the Web data (i.e., structured and unstructured, human-contributed and machine-generated), the fast expansion of information and the lack of standard information representation increase the complexity of knowledge extraction. The challenge we face is to identify relations from various of natural language forms or extract natural language forms from non-natural language expressions [18].

In this chapter, we give an introduction to the research fields related to our works of knowledge base management to help readers gain a better understanding of the works described in this book. In particular, we overview the techniques in knowledge extraction, including the extraction techniques and representation models. We also overview the querying techniques used for querying knowledge bases. We propose an architecture of knowledge management and address the main research issues, followed by the discussion of our contributions.

1.1 Overview of Knowledge Bases

The knowledge base is originated from the expert system or called knowledge-based system which was first developed by Artificial Intelligence (AI) researchers. A knowledge-based system consists of two components: a knowledge base and an inference engine. The knowledge base represents facts about the world. The inference engine represents logical assertions and conditions about the world [19]. The first knowledge-based systems represented facts about the world as simple assertions in a flat database and used rules to reason these assertions. With the evolution of database systems and information extraction techniques, knowledge bases become larger and more dynamic, being able to manage both structured and non-structured data in various domains. For example, WordNet [9] is a knowledge base for English lexicon. UniProtKB [20] is the knowledge hub for the collection of protein sequence and functional information. It also provides rich and accurate annotations by experts. GeoSpecies Knowledge Base [21] contains information on biological orders, families, species as well as species occurrence records and related data. NALT [22] is the National Agricultural Library Thesaurus, which holds vocabulary of agricultural terms in English and Spanish. There are also KBs that provide general information, such as DBpedia [5], Freebase [23], OpenCyc [24], Wikidata [7] and YAGO [6].

Knowledge bases can be categorized into curated KB and open KB [25] in terms of the way the knowledge is extracted and represented. Curated KBs [5–7, 23, 24] are built from collaboratively and manually constructed web corpus (e.g., https://en. wikipedia.org.) and the factual knowledge is represented using entity-relationship models. These knowledge bases have high data quality but suffer from low coverage, high cost for construction and fast ageing [6]. On the other hand, Open KBs [25–28] are proposed that automatically collect facts from web pages and construct assertions to represent knowledge. These KBs have higher coverage and up-to-date information but lower data quality compared to manual effort involved curated KBs. Moreover, they typically do not have explicit knowledge representation model and well-structured schema, which makes the queries restricted to only simple ones [25], although enabling more flexible knowledge extraction and representation.

Knowledge base is useful in application domains such as data integration, Named Entity Recognition (NER), topic detection and document ranking [5]. It is also adopted by search engines (e.g., Google) or question-answering systems (e.g., IBM's DeepQA project [29]) to facilitate the understanding of users questions. Knowledge base research is also popular in Semantic Web community because knowledge base can provide facts and relationships as well as support inference. Semantic Web, also has the name of Web 3.0 which is the next major evolution in connecting information. It enables data to be linked between sources and to be understood by computers so that they can perform sophisticated tasks on the behalf of human beings.[1] Linked Data[2] is a project to link data on the Web. It actually connects the public knowledge bases.

Research topics of knowledge base mainly include: (1) KB construction: information extraction (e.g., [30]), entity disambiguation (e.g., [31]), multilingual translation (e.g., [32]) and knowledge representation (e.g., [33]); (2) KB completion:

---

[1]http://www.cambridgesemantics.com/semantic-university/.
[2]http://linkeddata.org/.

entity linking (e.g., [34]) and fact inference (e.g., [35]); (3) KB usage: semantic enriched searching (e.g., [36]) and KB-Question Answering (KB-QA) (e.g., [25]). Our works presented in this book focus on the querying and information extraction in general KBs, but the techniques used can be applied to domain-specific KBs. In the following sections, we overview these techniques used in both curated KBs and open KBs.

1.2   Overview of Knowledge Extraction in Knowledge Bases

In this section, we overview the techniques involved in knowledge extraction (Sect. 1.2.1). We further introduce how the knowledge are represented in KBs (Sect. 1.2.2).

1.2.1   Extraction Techniques Overview

We discuss the knowledge extraction techniques in curated KB and open KB respectively.

1.2.1.1   Knowledge Extraction in Curated KBs

The knowledge extraction for curated KBs generally follows a process that includes four phases: Data acquisition, Facts extraction, Translation, Integration as shown in Fig. 1.1.

- *Data Acquisition Phase.* In Data acquisition phase, information is obtained from knowledge corpus using data dump or through APIs. Current curated KBs mostly depend on Wikipedia, one of the most popular online encyclopedias. For example, DBPedia obtain pages from Wikipedia through its data dump and from the MediaWiki API [5]. It extracts structured information such as from infobox tables, categorization information, geo-coordinates, and external links [1]. YAGO comprises information extracted from multiple resource: Wikipedia, WordNet and GeoNames and it also use data dump from these resources. Different from above KBs, Wikidata provides support for Wikipedia. It is a community effort that builds KB through crowd sourcing: information are manually added and edited, schema is maintained through community agreements [7].
- *Facts Extraction Phase.* In general, facts extraction includes entity determination, relation extraction, ontology mapping and checking/verification. The Wikipedia-driven KBs regard each page is an entity [5, 6, 37]. Then the relations (or facts) are extracted from infobox tables of Wikipedia pages as the infoboxes contain rich knowledge in the structural form of attribute-value pairs. The extracted

Fig. 1.1 Knowledge
extraction phases in curated
knowledge bases

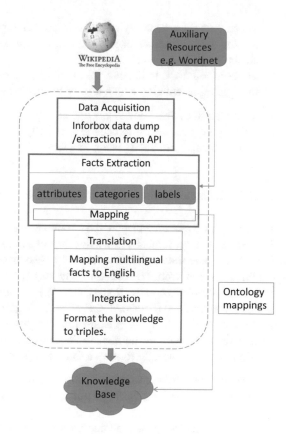

entities and relations then are mapped to the existing ontology of current KB
to form consistent knowledge representation. The final step is to do some
verification of the extracted entity-relationships, such as type checking [37]. The
key component in this phase is the relation extraction, which leverages traditional
information extraction techniques, but benefit from the structured Wikipedia
infobox information. The works that follow a way that take the known relations
(in training data) as seeds and identify relation patterns from them, then apply
these patterns to new corpus to find new relations. The techniques include entity
recognition, pattern matching, natural-language parsing, and statistical learning.
For a more comprehensive literature review, please refer to [38].

- *Translation Phase*. The translation in knowledge extraction is also called multilingual mapping or cross-lingual mapping. As English knowledge is much more
than other languages, knowledge in English is usually regarded as the base of
multilingual mapping, which means entities and attributes in other languages
are mapped to English entities and attributes. Many works are [39–42] about
consolidating information across multilingual Wikipedias. These works map
entities using inter-languages links [39, 40] or leverage external KBs [37] to get
the translation of entities. To map attributes, probabilistic methods are proposed

to estimate the possibility of multi-lingual mapping [37, 42]. Another way is to use the instance values of attributes to measure the similarities of multi-lingual attributes [41].

- *Integration Phase*. In the integration phase, the extracted knowledge is formatted to triple representation which will be discussed in Sect. 1.2.2.

In a curated KB, the knowledge update depends on the update of the resource knowledge corpus, which is usually manually created. Therefore, the knowledge extraction from curated KBs is not in a real time manner.

1.2.1.2   Knowledge Extraction in Open KBs

Knowledge extraction in open KBs focuses on the relation extraction. As open KBs do not have logical knowledge representation model, they cannot provide inferences. The data resources could be web pages, articles and any other data available on the Web that contain relational phrases. The relation extraction lies in open information extraction (Open IE) techniques, which extract relational tuples (or assertions) (*Argument1*, *Predicate*, *Argument2*) from large corpus without training data that contains existing relation patterns. Open IE has the ability to extract relations with no number limitation–it can extract as much as relations as possible. Open IE usually follows a four-step process (adapted from [43]):

- Parsing. Sentences are automatically parsed and labelled using natural language processing tools, e.g., POS (part-of-speech) tags and Noun Phrases (NP) chunking.
- Sequencing. The parsed sentences are split into clauses that contain relation phrases.
- Extractor Learning. It includes argument learning and relation phrase learning. The relation phrases are learned based on the sequences obtained from sequencing. Lexical constraint and syntactic constraint filter the relation phrases. The argument are learned through auxiliary training data or corpus.
- Extraction. Take the sentence as input, the words between the identified arguments are labelled as belong to a relation or not using learned extractors. Finally it outputs the relations to n-ary assertions.

Figure 1.2 illustrates these main phases of extracting facts for open knowledge bases.

Many research efforts are contributed to the open information extraction, aiming at building knowledge bases from web resources automatically. The work in [44] uses a Naive Bayes model and NLP features such as unlexicalized POS and NP-chunk features to train from the examples generated from Penn Treebank. The work in [45] follows this work and improves the extraction by leveraging Markov Logic Network. Some works also use Wikipedia as a auxiliary source to further improve the extractions, e,g, [46]. A recent work presented in [43] proposes a new way for

Fig. 1.2 Knowledge
extraction phases in open
knowledge bases

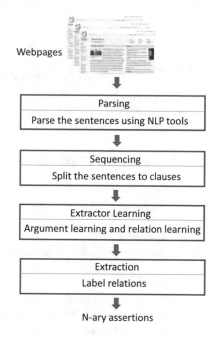

Webpages

Parsing
Parse the sentences using NLP tools

Sequencing
Split the sentences to clauses

Extractor Learning
Argument learning and relation learning

Extraction
Label relations

N-ary assertions

open information extraction by turning argument-centred IE to relation-centred IE
and achieve doubled precision/recall compared to previous Open IE methods.

1.2.2   Representation Models Overview

In KBs, the representation of knowledge has two components: data representations,
which describe how the facts are represented, and knowledge representation, which
defines the classes, categories and taxonomies for understanding the knowledge. We
overview the representation models of the two components in this section.

1.2.2.1   Data Representation

To represent facts, in curated KB, information is modelled as entities and relations
that strictly follow a predefined schema. Resource Description Framework (RDF)
is widely used as data modelling language for curated knowledge bases. RDF
represents a relationship by a three-element tuple, i.e., triple (*subject*, *predicate*,
object) in which *subject* and *object* are connected by *predicate*. A knowledge base
can be represented by a set of triples which are connected to form a graph (that is, a
RDF graph). So knowledge base modelled by RDF can also be called knowledge
graph. Moreover, RDF allows the sharing and reuse of data across boundaries.

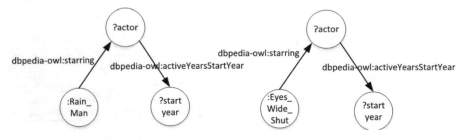

Fig. 1.3 Two examples of RDF subgraph

Sentence:
The U.S. president Barack Obama gave his speech on Tuesday to thousands of people.

Assertion 1:
(Barack Obama, is the president of, the U.S.)

Assertion 2-4:
(Barack Obama, gave, his speech)
(Barack Obama, gave his speech, on Tuesday)
(Barack Obama, gave his speech, to thousands of people)

Assertion 5:
(Barack Obama, gave, [his speech, on Tuesday, to thousands of people])

Fig. 1.4 Examples of assertions

Figure 1.3 shows two example RDF subgraphs and both of them illustrates two relationships (movie, starring, actor) and (actor, start year, year). To represent the RDF triples, several human readable formats are used, e.g. N-Triples[3] Notation 3,[4] and Turtle.[5]

In open KB, the facts are also represented by relations, but the representation does not follow a strict structured schema. A relation can be represented using a three-element triple or with additional prepositional phrases. These tuples are also called assertions. Figure 1.4 gives some examples of assertions obtained with Open IE technique [47].[6] Assertion 1–5 are extracted from the sentence on the top. Assertion 5 is the n-ary relationship version of assertion 2–4.

---

[3]https://www.w3.org/TR/n-triples/.
[4]https://www.w3.org/DesignIssues/Notation3.
[5]https://www.w3.org/TR/turtle/.
[6]https://github.com/allenai/openie-standalone.

1.2.2.2   Knowledge Representations

As Open KB does not have logical knowledge representation, we only introduce knowledge representation in curated KBs. Two methods are used in representing knowledge in curated KBs, namely ontology and rules. Although our works in this book do not cover either of the two methods, we briefly introduce them here for completeness.

Ontology formally defines the types, properties, and interrelationships of the entities that exist for a particular domain of discourse.[7] Each knowledge base can define its own ontologies or can use existing ontologies to describe taxonomies and the structure of knowledge. Ontology is one of the key components in curated KBs. To present ontologies, several ontology languages are proposed and several most used ones are introduced as follows:

- RDF Schema (RDFS) is a semantic extension of basic RDF data model. It defines a set of classes with certain properties using RDF. It describes the ontologies and is also called RDF vocabularies. RDFS can define entailment and such entailment can be used to retrieve answers that can not be explicitly obtained from KBs.
- Web Ontology Language (OWL) is a more expressive ontology language compared to RDFS and thus widely used. OWL has larger vocabulary to describe operations and relations and is more rigid to avoid ambiguity. It has more constraints than RDFS that enables variety of design of data management. It also provides a rich variety of annotations to enrich the reusability.
- Web Ontology Language for Web Services (OWL-S) is built upon OWL, which is an ontology for describing Semantic Web Services. It enables users and software agents to automatically discover and monitor Web services offered by Web resources. The OWL-S ontology has three main parts, namely the service profile, the process model and the grounding.
- Simple Knowledge Organization System (SKOS) can be regarded as a simplified and less formal ontology language compared to OWL and RDFS. It is proposed for applications with simple browsing or navigation interface. In this case, a completed and complex ontology language is not necessary. Moreover, SKOS is more human readable than the formal ontology languages.

Rules in curated knowledge bases define the way to discover and generate new relationships based on existing ones. Rules encode knowledge and simplify the first order logic which is the fundamental of building inference engines that can process conditions and output correct answers for queries. Many rule languages and systems have been proposed to offer varying features to reason about RDF data. Some examples are: Rule Markup Language (RuleML), which is a industry standards design to express rules and logics. And The Semantic Web Rule Language (SWRL), which is developed based on subset of RuleML and includes OWL DL or OWL Lite.

---

[7]https://en.wikipedia.org/wiki/Ontology_(information_science).

As there are not standard for rules languages, W3C develop a standard exchange format for rules: the Rule Interchange Format (RIF) to mitigate the situation.

1.3   Overview of Knowledge Bases Question Answering

The Knowledge Base Question Answering (KB-QA) is to retrieve answers of natural language questions from knowledge bases. The natural language questions are answered in a two-step way: Firstly, the free-form natural language question is transformed into a structured query which can be understood by the KBs; Secondly, the structured query is executed on the KBs and the answers are returned. KB-QA remains a challenging topic for both curated KBs and Open KBs but the techniques used are different in these two types of KBs. We overview these techniques in this section separately.

1.3.1   Question Answering in Curated KBs

In this section, we first discuss the research efforts on transforming natural language questions to structured query languages (e.g., SPARQL). Then the structured languages for querying RDF modelled knowledge base are discussed. At last, we overview the storage and indexing techniques used for querying triple stores (i.e., systems to store the RDF modelled knowledge base) that are not backed by structured query languages. Figure 1.5 illustrates these key components of querying curated KBs and we discuss them in more detail in this section.

1.3.1.1   Mapping Natural Language Question to Structure Query

The structured query languages such as SQL and SPARQL require language-specific knowledge (e.g., knowing of language syntax and data schema of queried KB) to form queries with them. These presents usability constrains to the end-users. Many works put effort on translating user-issued keywords or natural language questions to the structured query languages to reduce the expertise requirement on users. Natural language questions or keywords are turned into SQL [48], SeREQL [49] and mostly SPARQL [50–54]. The general process of query generation in these works has three steps: (1) mapping the keywords/entities in natural language questions to the entities in the querying KBs; (2) building query graphs for the structured query language using the entities; and (3) ranking the constructed queries. SPARK [50] processes natural language keywords requests and output a ranked list of SPARQL queries. It uses probabilistic model to rank the constructed queries. AutoSPARQL [51] uses supervised learning to generate SPARQL queries based on positive examples obtained from interactions with users (i.e., users estimate

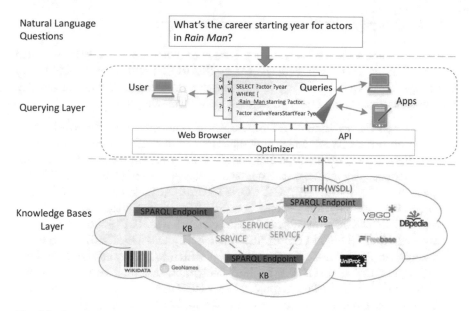

Fig. 1.5 Querying curated knowledge bases

whether an entity belongs to the expected answer set). The human judgement helps improve the interpretation. The work in [52] uses a template approach that determines SPARQL template from the syntactic structure of the natural language question. Query patterns are exploit in the work of [55] that the interpretation process is guided using predefined query patterns which represent the variations of most real queries. The approach avoids the entity mapping. DEANNA [53] focuses on solving disambiguation issue in transforming natural language to SPAROL by encoding the tasks into a integer linear program (ILP). It also leverages the external KB for identifying entities and relations. Very recently, authors of [54] propose a whole graph data-driven approach that models the natural language question in a structural way and reduces the question-answering on RDF data as a subgraph matching problem.

Ontology-based methods translate keywords or natural language to generate SPARQL queries using query ontologies. A general approach to ontology-based query interpretation follows two steps [56]: (1) Transforming natural language question to logical query under consideration of a lexicon and an underlying KB; (2) Transforming logical query into a query under consideration of the relations and topology in the KB. In Pythia [57], questions are directly mapped onto the vocabulary of the underlying ontology. FREyA [58] is a system that combines ontology-based lookup with syntactical parsing to interpret a natural language question. It also involves the interaction to users.

1.3.1.2 Querying with Structured Query Languages

There are several structured languages that have been developed for RDF modelled curated KBs and the most widely used is SPARQL Protocol and RDF Query Language (SPARQL). We briefly introduce these structured query languages here and then introduce using SPARQL Enpoint to perform querying.

- **Structured Query Languages.** SPARQLwas standardized by the World Wide Web Consortium (W3C) and is now supported by most RDF triple stores [59]. It is a SQL-like query language and allows a query to consist of triple patterns, conjunctions, disjunctions, and optional patterns. Different to SQL, SARPQ query has four forms: SELECT, CONSTRUCT, ASK and DESCRIBE where SELECT is used to extract values from SPARQL endpoint (like SELECT from SQL), CONSTRUCT transforms the extracted value to RDF graph, ASK returns boolean result (i.e., TRUE/FALSE) for a query and DESCRIBE is used to describe an RDF. All the four forms use WHERE to constraint the query [59]. SPARQL query is essentially based on subgraph matching as the query and RDF data can be modelled as graphs. Other RDF query languages include DQL (XML-based), queries and results expressed in DAML+OIL; N3QL (based on Notation 3); R-DEVICE; RDFQ (XML-based); RDQ (SQL-like); RDQL (SQL-like); RQL/RVL (SQL-like); SeRQL (SQL-like, similar to RQL/RVL) etc. [60].
- **SPARQL Endpoint.** A SPARQL endpoint is a SPARQL protocol service defined in the SPROT specification [61]. A SPARQL endpoint can be regarded as the querying interface towards a knowledge base as it enables users to query a knowledge base via the SPARQL language and without knowing the underlying data schema. In this way, querying is like performing queries on a search engine through a Web interface. This is realized by the HTTP bindings provided by KBs, which is written in Web Services Description Language (WSDL). For federated queries, which aim to access data across SPARQL endpoints, SPARQL 1.1 specification introduces the SERVICE keyword for this purpose. The querying layer and knowledge base layer in Fig. 1.5 illustrate how the HTTP bindings and SERVICE keyword enable querying on underlying curated KBs.

1.3.1.3 Storage and Indexing in Underlying Databases

The underlying systems of curated knowledge bases are databases, thus the techniques of querying curated KBs are mainly developed from the database community. Traditional RDBMS systems are first applied to store the RDF graphs, with the development of database systems, graph databases are introduced. In a system perspective, distributed databases are developed to store larger RDF graph than single machine based databases. Moreover, in-memory databases are used to provide efficient querying functionality. In this section, we overview the databases that are used to store the RDF graphs and introduce the indexing techniques of them as the querying efficiency are highly depend on the storage and indexing of these databases.

- **Storage**. RDBMSs have been developed for dozens of years and they are very mature with desirable performance and many functional and optimization tools. It is natural for many researchers to choose RDBMSs as RDF storage systems. In many existing systems, the triples are directly stored in relational tables and the method can be further classified into three ways: store RDF as triple table (e.g, Sesame [62], RDF-3X [63]); as property tables (e.g, Jena [64], BitMat [65, 66]), and as vertical partitioned tables (e.g, C-Store [67], SW-Store [68] and [69]). Graph-based approaches aim at improving the performance of graph-based manipulations on RDF datasets by leveraging the nature format of RDF graph. In [70], a method stores RDF graph as an object in object-oriented database is introduced. GStore [71] encodes all vertices and edges and transforms the sub-graph matching issue to bit operation. Trinity.RDF [72] represents the graph with adjacency lists and combines traditional query processing with graph exploration. And it is a distributed system. There are other distributed systems such as SHARD [73], YARS2 [69], and Virtuoso [74] introduced to store larger RDF graphs. In terms of disk-based or in-memory, most of the systems mentioned previously are disk based, while recently, some in-memory systems are introduced in order to reduce the huge overhead of disk I/O. For example, RDF-3x [63], Trinity.RDF [72] and BitMat [66] are in-memory systems.
- **Indexing**. Indexing is closely related to storage and query. Many systems enhance their systems by introducing some modifications on indexes aiming to fulfil scalability requirement to a certain extent. RStar [75] is the first proposal for B-Tree indexes on the triple table. It introduced unclustered indexes on S, O, P separately and combined indexes on (SP), (PO) and (SPO). Virtuoso [76] stores different permutations of SPOC, where C indicates the Context column. Also Virtuoso builds bitmap indexes for each OPC prefix by default. RDF-3X, Hexastore etc. propose exhaustive indexes of triples and all their embedded sub-triples (pairs and single values) in six separate B+-tree or hash indexes. In [77], a tree-shaped indexing structure called GRIN index is proposed where each node in the tree identifies a sub-graph of the indexed RDF graph. GRIN shows good performance on small- to medium-sized data and for hand-specified execution plans [63]. [78, 79] support literal-specific indexes by introducing inverted indexes on string literals. BitMat [65] builds a virtual 3D bitmap upon an RDF graph, with each dimension representing the subjects, predicates and objects, respectively. TripleBit [80] develops two auxiliary indexing structures, ID-Chunk bit matrix and ID-Predicate bit matrix, to reduce the number and the size of indexes to the minimum while providing orders of magnitude speed up for scan and merge-join performance. gStore [71] builds indexes over signatures using VS-tree. The signature is a bit-string that contains some information about the triples in which the resource occurs as the subject. Some research introduced in-memory indexes in RDF systems: [81] proposes three hash-base in-memory indexes. Kowari [82] builds six sparse indexes on SPOC using AVL trees. YARS2 [83] considers the use of extensible hash-tables and in-memory sparse indexes instead of B-Trees by mimicing six B-Trees using sixteen hash-tables.

1.3.2   Question Answering in Open KBs

The question-answering (QA) techniques in Open KBs are derived from information retrieval field, hence very different from the techniques used in QA in Curated KBs. Although the Open IE is not a new technique, QA on Open KB is relatively new. We discuss several recent works in this section as an overview of question answering in Open KBs.

PARALEX [27] learns to map questions to formal queries directly by performing paraphrasing, parsing and query-rewriting in one step. Answers are then returned by executing these formal queries. The learning algorithm is similar to algorithms used for paraphrase extraction from sentence-aligned corpora. Specifically, PARALEX first uses 16 seed question templates to find high-quality queries. Based on the question-query pairs, it learns a large set of lexical equivalence by leveraging word alignment. Then PARALEX uses a liner ranking model to filter learned lexical equivalence and keep the ones that with highest ranking score as the answer. As the Open KB contains only assertions that restrict the QA available to simple questions, TAQA [25] provides a method to answer complex questions against the KBs with n-ray assertions that contain more complex information, e.g., prepositional or adverbial constraints. In order to get the answer with these constraints, the work builds a KB that contains assertions with an arbitrary number of arguments. For QA, TAQA divides the process into several sub-problems: question paraphrasing, question parsing, Open KB querying and answer ranking. OQA [28] decomposes the QA into smaller sub-problems and solves these sub-problems by mining large rule set from question corpus and KBs–both Curated KBs and Open KBs. For question paraphrasing sub-problem, OQA mines source/target template pairs from WikiAnswers. The paraphrasing operators are then generated from the most co-occurring pairs. For question parsing sub-problem, OQA uses 10 hand-written templates to map questions to queries. For query rewriting, OQA mines query rewrite rules from multiple KBs. The final query execution is performed on inverted index on the extracted 3-tuple assertions from KBs. A very recent work [36] also uses Curated KB as auxiliary resource to assist the QA on Open KB. Specifically, it links the answer candidate to entities in Freebase to reduce the redundancy of answer candidates and automatically obtain the answer type, which leads higher accuracy in QA.

1.4   Research Issues in Querying and Extracting Knowledge Bases

The works in this book tackle a number of research issues in querying and knowledge extraction in knowledge bases. Although different chapters address different tasks, we use this motivating scenario as a generic example.

1.4.1   An Architecture for Knowledge Base Management

Figure 1.6 illustrates a management architecture that includes the knowledge extraction and querying of knowledge bases. Knowledge consumers are on the top of the figure. Not only human can consume the knowledges, machines such as computer programs can understand and use the machine-understandable knowledge. The interfaces or API prepare the querying layer between knowledge consumers and the knowledge bases. This layer contains the query scheduling, optimization and provides the protocols that enables the querying against knowledge bases. In the middle of the figure lies the knowledge bases which is the center of this motiving scenario. Knowledge bases can also be linked together to provide a large network to share knowledge. In the knowledge extraction layer, the data turns into knowledge by following four steps: data acquisition, data cleaning, data clustering and knowledge extraction. Data acquisition takes the responsibility to crawl data

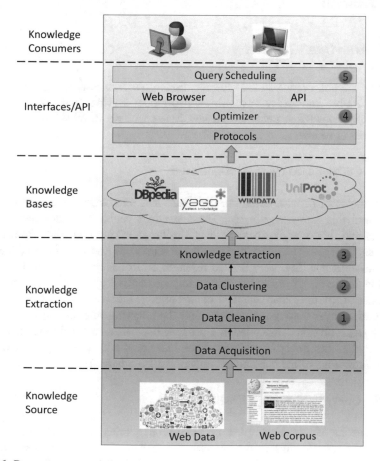

Fig. 1.6 Data management in knowledge bases

from the Web or extract data from structured Web corpus. Data cleaning covers the tasks of pruning invalid data, reducing redundancy, disambiguation etc. In data clustering, data are grouped to different domains and input the clustered data into knowledge extractor, which finally extract the relations/facts/knowledge.

This motivating scenario poses several research directions including: (1) Guaranteeing the quality of the data as input to the extractor due to the existence o high redundancy and low quality data (Fig. 1.6, issue 1); (2) Clustering the knowledge to the correct domain during the automatic extraction process as for domain-specific knowledge bases, cross-domain information would mislead knowledge consumers (Fig. 1.6, issue 2); (3) Extracting knowledge from heterogeneous sources, including structured and unstructured, natural language and non-natural language resources (Fig. 1.6, issue 3); (4) Optimizing queries for more efficient knowledge retrieval, especially in the case that consumers use the interface to perform the querying (Fig. 1.6, issue 4); (5) Predicting the performance of subsequent queries for appropriately query scheduling to avoid long-running queries occupy the resources (Fig. 1.6, issue 5).

1.4.2 Our Contributions

Based on the observation in the aforementioned research issues, querying against public available knowledge bases through protocol services and efficient extract high quality knowledge raise several key issues and we provide the solution for these issues as follows:

1.4.2.1 Cache-Based Query Optimization

Knowledge bases provide certain protocol services for consumers to query against public available knowledge bases. These services enable the querying from an interface and this querying behaviour is similar to asking questions from a Web search engine. However, querying using these services is known to have latency due to the network inability. Very limited research effort has been put into improving this kind of querying from a client-side perspective as very limited information can be obtained from the client-side. Our contribution is that we propose a cache-based optimization layer between consumers and the querying interface to facilitate the querying and solve the latency issue. The cache depends on a novel learning method that considers the querying patterns from individual's historical queries without having knowledge of the backing systems of the knowledge base.

1.4.2.2 Query Performance Prediction

Consumers could issue complex or erroneous queries that could occupy the system resources, leading to the hanging of other queries. Effectively predict the performance of the queries is the key component of query scheduling, in which the estimated long-running queries could be rescheduled when the resource is not in contention. We work on the prediction of structured queries on public available knowledge bases. As the structure queries are complex and understandable only to people knowing the syntax of the querying language and the schema of the underlying database, it is difficult for novice data consumers to obtain the features of the queries—the features are used to feed the prediction model. Our contribution is to eschew the knowledge requirement on both the querying languages and system. Instead, we obtain features from syntactic and structure characteristics of the queries and use these features as well as the performance metrics to predict the performance of subsequent queries.

1.4.2.3 Data Cleaning for Knowledge Extraction: Paraphrase Identification

With the booming of data on the Web, more knowledge are available from hetero-geneous data sources. Human contributed data always occupy a large proportion of the data, but have problems in data quality. For example, erroneous data could be provided by mistake and duplicate information exist due to the multi-sources. The data could contain lexical and syntactical incorrect phrases and human-understandable abbreviations. This requires the involvement of natural language processing techniques in the data cleaning stage, preparing data for the follow-up knowledge extraction. Our contribution lies on the proposed paraphrase or duplicate identification method that focus on the redundancy issue of data cleaning. To tackle this issue, we develop novel features that consider both textual and latent characteristics of questions from question-answering communities, in which the questions are created by non-expert people and contain erroneous phrases.

1.4.2.4 Data Clustering for Domain Specific Knowledge Extraction

Data in multiple domains widely exist on the Web. Building domain-specific knowledge bases from resources that contain multi-domain knowledge require techniques to effectively separate knowledge into different groups. This step makes the knowledge extraction easier as for example, ambiguous entities in different domains will not exist in domain-specific groups. To tackle this problem, we use clustering-based method to cluster the knowledge. Our contribution is that we develop a novel clustering algorithm that achieves both accuracy and efficiency.

1.4.2.5 Knowledge Extraction from Unstructured Data on the Web

Unstructured data can be both textual (e.g., email message, tweets and source codes) and non-textual (e.g., images, audio and video files). It becomes of huge volume as it is generated daily in this information era. It is challenging to automatically extract knowledge from unstructured data because the data lack of schema and could not be easily understood by machines. Moreover, it is usually with low data quality. We focus on the knowledge extraction from one type of unstructured data on the Web: the source codes from large Web repositories. The source codes are not only unstructured, but also typically non-natural language, which raise the complexity of the extraction. Our contribution is to leverage natural language processing techniques to pre-process the source codes and obtain the natural language elements from source code. Based on the extracted natural language texts, we apply traditional knowledge extraction techniques.

1.5 Book Organization

The reminder of this book is organized as follows:

In Chap. 2, we present our work on predicting performances of SPARQL queries in a client-side perspective. We first detail our feature modelling approach and then apply several state-of-the-art prediction models on these features. We also discuss some important observations from our extensive experiments.

In Chap. 3, we discuss our learning-based cache framework proposed for solving the query latency from SPARQL endpoints. Similar to Chap. 3, the feature modelling is the key contribution of this effort and we detail our novel approach to identify the queries to be cached and a novel adaptation of a cache replacement algorithm.

In Chap. 4, we describe our proposed clustering algorithm aiming at the task of knowledge clustering. We model the clustering problem as a matrix factorization problem, which leverages optimization approach to get the optimal solution. The clustering constrains are realized on the constraints from manifold which makes the algorithm more fast and accurate.

In Chap. 5, we describe the topic extraction method aiming to extract knowledge from the unstructured data, particularly the source codes. We first discuss our effort on pre-processing the untidy data. Then we present the method we develop for topic extraction. A novel automatic term selection algorithm is also introduced and this algorithm helps identify the most contributory terms for the topic extraction performances.

In Chap. 6, we first introduce our proposed prototype for building knowledge base from programming question answering communities. Then we focus on the discussion of duplicate posts detection method, which is the first step for preparing data to build such a knowledge base. We also report our evaluation on multiple datasets in various domains.

Finally, in Chap. 7, we provide concluding remarks of this book and discuss future work directions.

Chapter 2
Cache Based Optimization for Querying Curated Knowledge Bases

In this chapter, we introduce our proposed cache based optimization framework to improve the overall querying performance on the SPARQL endpoints that are built upon knowledge bases. The framework is named SPARQL Endpoint Cache Framework, SECF, which utilizes machine learning techniques to learn clients' query patterns and suggests similar queries, whose results are prefetched and cached in order to reduce the overall querying time. We also developed a cache replacement algorithm, Modified Simple Exponential Smoothing, MSES, to replace the less valuable cache content, i.e., queries, with more valuable ones. MSES outperforms the most used cache replacement algorithm LRU-2. Our approach has been evaluated by using a very large set of real world queries. The empirical results show that our approach has great potential to enhance the cache hit rate and accelerate the querying speed on SPARQL endpoints. This chapter is based on our research reported in [84–86].

2.1 Design Overview

Knowledge Bases (KBs) are widely used as one of the fundamental components in Semantic Web applications as they provide facts and relationships that can be automatically understood by machines (e.g., computer programs). Knowledge bases can be categorized into curated KBs and open KBs as discussed in Chap. 2.

In this chapter, we discuss the querying problem in curated KBs and we will use KBs and curated KBs interchangeably hereafter. The querying on curated KBs are realized by the HTTP bindings provided by KBs, for example, the SPARQL protocol, which is written in Web Services Description Language (WSDL[1]). In this way, clients do not need to access the background databases. Instead, querying

---

[1]https://www.w3.org/TR/rdf-sparql-protocol/.

© Springer International Publishing AG, part of Springer Nature 2018
W. E. Zhang, Q. Z. Sheng, *Managing Data From Knowledge Bases:
Querying and Extraction*, https://doi.org/10.1007/978-3-319-94935-2_2

is like performing queries on a search engine through a Web interface. For federated queries, which aim to access data across SPARQL endpoints, SPARQL 1.1 specification introduces the SERVICE keyword for this purpose.

Currently, querying SPARQL endpoints has the problems like network instability and latency, which affect the query efficiency. Therefore, the most typical way for consumers who want to query public data is downloading the data dump and setting up their own local SPRAQL endpoint. But data in a local endpoint is not up-to-date and hosting an endpoint requires expensive infrastructural support. Many research efforts have been dedicated to circumvent this problem [14–16, 87, 88] and caching is one of the popular directions [89]. Caching in search engines facilitates query processing in terms of both latency reduction (the request is answered from the cache, which is closer to the client) and network traffic reduction (much information requested is reused). Likewise, it can be leveraged to accelerate query response speed on SPARQL endpoints. While most research efforts focus on providing a server-side caching mechanism, being embedded in triple stores (e.g., Jena[2] and Joseki[3]), client-side caching has not been fully explored [14]. Server-side caching is usually embedded in the background databases. Sometimes they are part of, or cooperate with, the query optimizer. Server-side cache is well developed but it is not customized to catch the different querying patterns from clients. Moreover, the design and development of server-side cache highly depend on the knowledge of background databases/servers. It is not easy to develop a generic approach.

On the other hand, "client-side caching" is a terminology from Web techniques, where the background databases/servers are black boxes to the Web users. Using client-side caching avoids making repeated request to the servers and can quickly get answers. In addition, it is possible to collect client's querying behaviours.

Our approach, SPARQL Endpoint Caching Framework (SECF), adopts the client-side caching idea and is a domain-independent client-side caching framework. SECF caches the (query, result) pairs for current processing query and its similar queries. This is motivated by the observation that end users who consume RDF-modelled knowledge typically use programmatic query clients, e.g., software or services to retrieve information from SPARQL endpoints [87]. These queries usually have repetitive query patterns and only differ in specific elements of a triple pattern (a triple pattern is similar to a triple, except that at least one element namely subject, predicate or object, is a variable). Moreover, they are usually issued subsequently. To illustrate, Fig. 2.1 gives two example queries that are structurally similar. Query 1 retrieves start year (i.e., the year their acting careers started) from the actors of the movie *Rain Man* and the year should be later than 1980. Query 2 requests the same information but for a different movie (*Eyes Wide Shut*). The differences between these two queries are the movie names (the underlined terms), which is the subject element of triple pattern "*movie* dbpedia-owl:starring ?actor" and the year in the Filter expression. By considering these observations, we propose

---

[2]http://jena.sourceforge.net/.
[3]http://sourceforge.net/projects/joseki/.

```
Query 1
SELECT ?actor ?year WHERE {
+underline[:Rain_Man] dbpedia-owl:starring ?actor .
?actor dbpedia-owl:activeYearsStartYear ?year .
}
FILTER(?year>1980)

Query 2:
SELECT ?actor ?year WHERE {
+underline[:Eyes_Wide_Shut] dbpedia-owl:starring ?actor .
?actor dbpedia-owl:activeYearsStartYear ?year .
}
FILTER(?year > 1960)
```

Fig. 2.1 Example of similar queries. The queries only differ in the movie name and year

to prefetch and cache the query results of similar queries in advance. Since these queries are potentially subsequent queries, the average query response time will be reduced if the subsequent queries are already in the cache (cache hit) because the results are returned immediately rather than being retrieved from SPARQL endpoints.

The problem then turns into how to find similar queries that are potential subsequent query. To this end, SECF utilizes machine learning techniques to learn from the historical queries and captures the querying characteristics of the users. Then SECF suggests similar queries for a new query through the learned model. The key challenge centres on how to transform queries to vector representation that can be used by learning algorithms. We adopt the idea introduced in [90] to measure similarity between queries by leveraging *Graph Edit Distance* (GED) between the query patterns of queries. Specifically, it calculates the GEDs between the Basic Graph Patterns (BGPs) of all training queries, and then clusters the queries based on these GEDs. A feature vector of a query is constructed using the GEDs between this query and all clusters' centroid queries. So the number of features of a query equals to the number of clusters. We name this method *Clustering-based feature modelling*. However, there are two problems of clustering-based feature modelling. First, it only considers BGPs but ignores other most used SPAQRL operators which cannot be decomposed to BGPs, e.g., FILTER, BIND and VALUES. These operators, especially FILTER, occur frequently in real world queries (see Sect. 2.3.2 for our analysis on a large set of real-world SPARQL queries). Second, it requires the calculation of GEDs between all training queries which is time-consuming when the training size is large. We propose *Template-based feature modelling* to address these two issues. In template-based feature modelling, we consider operators FILTER, BIND and VALUES. We choose representative queries as target queries and calculate the GEDs between a query and these target queries. These GEDs are used to construct feature vectors. Template-based feature modelling drastically reduces the computation time compared to the clustering-based feature modelling presented in [90].

SECF then adopts k-Nearest Neighbour (k-NN) model to learn from the feature vectors of training queries and to suggest similar queries of a new issued query Q. k-NN [91] is originally a classification algorithm, that requires labels of training data. We modify it to an unsupervised learning algorithm that builds a k-dimension tree (k-d tree) according to the distances (*Euclidean* distance is used here) between feature vectors of SPARQL queries and does not require labels, which we don't have. For Q, SECF searches the k-d tree and finds its k-nearest neighbours, which are considered as the k most similar queries of Q. The suggestion process runs in a background thread to the query process. The training and mining process can be performed once as a pre-computing step. During the run-time, this background approach will give suggestions for similar queries based on the queries it has already processed. As queries that are never seen will come, the training set will be updated periodically to reflect the changes. This method adopts cache technique in database, where query plans are cached for exact the same queries performed later and the corresponding statistics are updated periodically. After identifying similar queries, SECF prefetches the results of these similar queries and caches the (query, result) pairs.

As the cache space is limited, less useful data should be removed from the cache. A cache replacement algorithm is introduced for this purpose. However, techniques for relational databases (e.g., LRU [92], LRU-k [93] and ARC [94]) cannot be directly applied into our client-side caching framework because our caching is record based, rather than traditional page-based caching algorithms. Moreover, our client-side application is not based on RDBMS and is not designed for server side as traditional caching algorithms do. In this work, we use time-aware frequency based algorithm, which leverages the idea of a novel approach recently proposed for caching in main memory databases in *Online Transaction Processing (OLTP)* systems [95]. More specifically, we use Modified Simple Exponential Smoothing (MSES) to evaluate the frequencies of cached queries and remove the ones with the lowest hit frequencies from the cache.

The contributions of this work are three folds. Firstly, we address the problem of providing client-side caching for accelerating query answering process for SPARQL endpoints and design a caching mechanism. Our mechanism can either be deployed as a Web browser plugin or be embedded in the firewall, but ultimately we envisage it being embedded within SPARQL endpoints that act as clients to other SPARQL endpoints by interpreting the SERVICE keyword for SPARQL 1.1 federated queries. Secondly, SECF suggests similar queries by leveraging machine learning techniques. The distance measurement for SPARQL queries considers both BGPs and the most used SPARQL operators. SECF also provides a smoothing-based cache replacement algorithm. Thirdly, we perform extensive experiments on real world queries to showcase the effectiveness of SECF.

The remainder of this chapter is structured as follows. We introduce SECF in Sect. 2.2. Then we report the experimental results Sect. 2.3. Finally. we give some discussions in Sect. 2.3.6 and overview the related work in Sect. 2.4.

2.2   The SPARQL Endpoint Cache Framework

Figure 2.2 illustrates the working processes of SECF. When a new query is issued, SECF first checks if query recording is enabled (①). If yes, a background process will log all queries into a file for further machine learning processing. Then it checks if an identical query (either cached as an issued query or a suggested query) has been cached (②). In this case, the results are returned immediately via the cache model (⑥ and ⑦). Otherwise, SECF fetches and returns the results directly from the SPARQL endpoint (③ and ④). These results are cached in the cache module (⑤). When query suggestion is enabled, during run-time, suggested queries are generated for the current query in suggestion module. The results of these suggested queries will be retrieved (⑧) from the SPARQL Endpoint in advance and cached (⑨), together with the queries in the form of (query, result) pairs (q_i, r_i) in ⑧. The aim of prefetching and caching similar queries in advance is to enhance the hit rate of cache (i.e., how much percentage of queries can be answered immediately from cached results). A cache replacement algorithm is executed when the cache is full or the number of cache queries is reached. It runs in a separate thread so that it does not affect the query answering process.

The overall query speed depends on the hit rate of the cache. Further, the number of queries that can contribute to hit rate can be improved by the process of prefetching and caching of similar queries to existing queries as it is observed from real world SPARQL queries that most subsequent queries are similar to previous issued queries [87]. If we cache the similar queries, which are potential subsequent

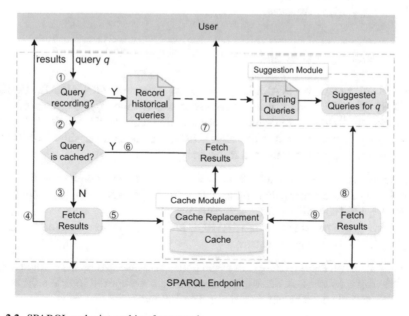

Fig. 2.2 SPARQL endpoint caching framework

queries, higher hit rate can be achieved. To identify and cache similar queries, we propose a learning based approach that consists of following three main steps:

- Step 1: Feature modelling. We propose to model SPARQL query to feature vectors that can be fed into multiple learning algorithms. However, how to model a SPARQL query in vector representation is a challenge problem. We first introduce the distance measurement between SPARQL queries in Sect. 2.2.1 and then discuss our feature modelling approach based on this distance in Sect. 2.2.2.
- Step 2: Training and suggestion. After obtaining the feature vectors of SPARQL queries, we train a suggestion model using historical queries as the training set. A trained model is the output. When a new query Q arrives, we first transform Q to a feature vector using techniques from Step 1. Then we feed the vector into the trained suggestion model for similar queries recommendation. We introduce our approach in Sect. 2.2.3.
- Step 3: Cache and replacement. We prefetch the results of similar queries. As the cache is with limited size, less useful queries (and their results) should be removed from the cache to enhance the hit rates of more frequently issued queries. We introduce our cache and replacement algorithm in Sect. 2.2.4.

2.2.1 Query Distance Calculation

Prior to discussing distance approaches, we briefly give definition of SPARQL queries and the Basic Graph Patterns (BGPs) and then introduce BPG distances and other operators' distances.

2.2.1.1 SPARQL Query and BGP

The official syntax of SPARQL1.1 considers operators OPTIONAL, UNION, FILTER, SELECT and concatenation via a dot symbol (.) to group patterns. VALUES and BIND are to define sets of variable bindings. We use B, I, L, V for denoting the (infinite) sets of blank nodes, IRIs, literals, and variables. A SPARQL graph pattern expression is defined recursively as follows [96]:

1. A valid triple pattern $T \in (IVB) \times (IV) \times (IVLB)$ is a graph pattern,
2. If P_1 and P_2 are graph patterns, then expressions $(P_1$ AND $P_2)$, $(P_1$ UNION $P_2)$ and $(P_1$ OPTIONAL $P_2)$ are graph patterns,
3. If P is a graph pattern and R is a SPARQL build-in condition, then the expression $(P$ FILTER $R)$ is a graph pattern.

A Basic Graph Pattern (BGP) is a graph pattern represented by the conjunction of multiple triple patterns. A SPARQL query can be decomposed to BGPs for certain operators and the decomposition is defined as follows:

Definition 2.1 (SPARQL Query Decomposition) Let $Q = (S_Q, P_Q)$ be the query where S_Q is the SELECT expression and $P_Q = P_1 \oplus \ldots \oplus P_n$ is the query pattern with $\oplus \in \{$AND, UNION, OPTIONAL, FILTER, BIND, VALUES, MINUS$\}$. When pattern feature $\oplus \in \{$AND, UNION, OPTIONAL, MINUS$\}$, graph pattern $P_i, i \in [1, n]$ can be recursively decomposed to sub-level graph patterns until the graph pattern is a BGP which can further be decomposed to triple patterns as $P_{bgp,i} = T_1 \oplus \ldots \oplus T_k$, where $\oplus = $ AND. When pattern feature $\oplus \in \{$FILTER, BIND, VALUES$\}$, graph pattern P_i cannot be decomposed to BGPs and is represented as expressions.

The decomposition of a SPARQL query is a recursive process with the result that can be regarded as a hierarchical tree. Further, it is easy to observe that query Q can also be represented as $Q = (S_Q, \{P_{bgp}, P_{filter}, P_{bind}, P_{value}\})$ where $P_{bgp}, P_{filter}, P_{bind}, P_{value}$ are BGP, FILTER, BIND and VALUE patterns in P_Q respectively. Note that each graph pattern can appear multiple times in a query pattern.

To find similar queries, we compute the distance between two given SPARQL queries by calculating the distance between patterns of the two queries:

$$d(P_Q, P_Q') = d(P_{bgp}, P_{bgp}') + d(P_{filter}, P_{filter}')$$

$$+d(P_{bind}, P_{bind}') + d(P_{value}, P_{value}') \tag{2.1}$$

where P_Q contains $P_{bgp}, P_{filter}, P_{bind}, P_{value}$ and P_Q' contains $P_{bgp}', P_{filter}', P_{bind}', P_{value}'$. $d(P_Q, P_Q') = 0$ denotes the two queries are structurally the same.

2.2.1.2 BGP Distance

We propose to use Graph Edit Distance [97] to measure the distance between BGPs as a BGP can be represented as a graph. GED between two graphs is the minimum amount of edit operations (i.e., deletion, insertion and substitutions of nodes and edges) needed to transform one graph to the other. To exemplify, Fig. 2.3 illustrates the graph representation of two triple patterns *(?s, p, o)* and *(?s, p, ?o)*. *s* denotes the *subject*, *p* denotes the *predicate* and *o* denotes the *object*. The question mark indicates that the corresponding component is a variable. However, it is hard to tell the differences of these two graphs, as they are structurally identical (i.e., GED between these two graphs equals to zero). Therefore, we formulate the problem of modelling BGPs to distinct graphs as follows:

Problem 2.1 (BGP Graph Modelling) Given $P_{bgp_i} = \{tp_1, tp_2, \ldots, tp_n\}$ denote a BGP of a SPARQL query, $tp_k, \ k \in (1, n)$ is a triple pattern rooted at P_{bgp_i}.

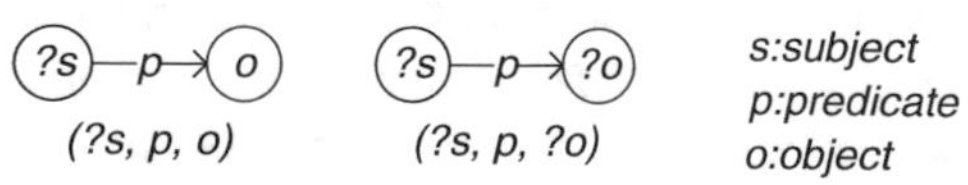

Fig. 2.3 Two example BGPs (?s, p, o) and (?s, p, ?o)

(?s, p, o)   (?s, p, ?o)

s:subject
p:predicate
o:object

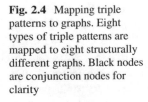

Fig. 2.4 Mapping triple patterns to graphs. Eight types of triple patterns are mapped to eight structurally different graphs. Black nodes are conjunction nodes for clarity

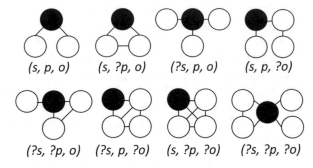

Fig. 2.5 Graph modelling for BGPs in Query 1

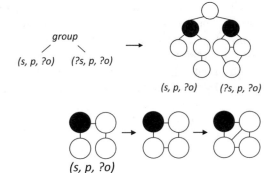

Fig. 2.6 Graph edit path from *(s, p, ?o)* to *(?s, p, ?o)*

$ged(g_o, g_d)$ represents the graph edit distance between graph g_o and graph g_d. BGP graph modelling is the task that models each tp_k to a graph g_{tp_k} satisfying $ged(g_{tp_k}, g_{tp_l}) > 0$ when $k \neq l$.

To address the above problem, we propose to map all the eight types of triple patterns to eight structurally different graphs, as shown in Fig. 2.4. The black circles denote conjunction nodes for clarity. They are not coloured in graph modelling. As we only consider the structures of queries, whether the connecting node represents a join or union is not distinguished. Therefore, the different meanings of connecting nodes are not considered in this work. Using these mappings, we model the triple patterns of BGPs in Query 1 in Fig. 2.1, to a graph, which is depicted in Fig. 2.5. The GED between *(s, p, ?o)* to *(?s, p, ?o)* in Fig. 2.4 is illustrated in Fig. 2.6.

There are various ways to map triple patterns to graphs. However, in our work, the way that we choose for the mapping does not affect the final cache result very much. So we only focus on mapping distinct triple patterns to distinct graphs. The reason is that in our work, similar queries that can lead to a cache hit are mostly the ones that are structurally the same with the current processing query. Specifically, the k-NN model first returns the structurally the same queries as similar queries, then returns the ones that are structurally the same but with different filter and bindings, and finally returns structurally similar queries. As the queries are mostly issued by programmatic clients and are generated with query templates (as described in Sect. 2.1), most returned queries are the ones using the same template, i.e., are structurally the same. Thus the structurally similar queries will not affect the caching results much as they are with small numbers. So the distances between different triple patterns that are considered in measuring structurally similar queries give limited impact on caching performance. Therefore, the difference of various mapping ways is not considered in this work.

2.2.1.3   Other Distances

We calculate other distances $d(P_{filter}, P'_{filter}), d(P_{bind}, P'_{bind})$ and $d(P_{value}, P'_{value})$ only when $d(P_{bgp}, P'_{bgp}) = 0$. We define distance between two FILTER expressions as half of their *levenshtein* distance when the variables in these two expressions are identical, otherwise the distance is a fixed value 1. Thus the distance is in the range of $[0, 0.5]$ or equals to 1.

$$d(P_{filter,i}, P'_{filter,i}) = \begin{cases} \dfrac{levenshtein(E(i),E'(i))}{2max(length(E(i)),length(E'(i)))}, \\ \qquad if \ \ V(i) = V'(i) \\ 1, \qquad else \end{cases} \qquad (2.2)$$

where $E(i)$ and $E'(i)$ represent the FILTER expression for $P_{filter,i}$ and $P'_{filter,i}$. $V(i)$ and $V'(i)$ are variables in these two FILTER patterns respectively. When there are multiple Filter expressions that can be compared, the total difference is defined as:

$$d(P_{filter}, P'_{filter}) = \sum_{i=1}^{m} d(P_{filter,i}, P'_{filter,i}) \qquad (2.3)$$

Filter expressions in Query 1 and Query 2 are similar as the distance is 0.05 using Eq. (2.3). So $d(P_{Q1}, P'_{Q2}) = 0.05$ (Eq. (2.1)).

We can also have similar functions for BIND and VALUE patterns.

$$d(P_{bind,i}, P'_{bind,i}) = \begin{cases} \dfrac{levenshtein(E(i),E'(i))}{2max(length(E(i)),length(E'(i)))}, \\ \qquad if \ \ V(i) = V'(i) \\ 1, \qquad else \end{cases} \qquad (2.4)$$

$$d(P_{bind}, P'_{bind}) = \sum_{i=1}^{m} d(P_{bind,i}, P'_{bind,i}) \qquad (2.5)$$

And

$$d(P_{values,i}, P'_{values,i}) = \begin{cases} \frac{levenshtein(E(i),E'(i))}{2max(length(E(i)),length(E'(i)))}, \\ \qquad if \ \ V(i) = V'(i) \\ 1, \qquad else \end{cases} \qquad (2.6)$$

$$d(P_{values}, P'_{values}) = \sum_{i=1}^{m} d(P_{values,i}, P'_{values,i}) \qquad (2.7)$$

2.2.2   Feature Modelling

Using the distance function Eq. (2.1), it is intuitive to suggest similar queries to a given query Q by calculating the distances between Q and each query in training set, then to rank the distance scores and find the top k similar ones. However, this method is time consuming as the calculation of distances between Q and each query in training set requires large amount of computation. Therefore, we choose to construct feature vectors for SPARQL queries that leverages the distances and can facilitate the similar queries suggestion. It is worth mentioning that the work in [90] proposes an approach to transform SPARQL query to vector representation. For comparison, we firstly introduce this approach, which we refer to as *cluster-based feature modelling* (Sect. 2.2.2.1) and then discuss our approach, the *template-based feature modelling* (Sect. 2.2.2.2).

2.2.2.1   Cluster-Based Feature Modelling

In cluster-based feature modelling, distances between each pair of queries in the training set are calculated using only BGP distance. Then *k-medoids* algorithm [98] is utilized to cluster the training queries by using distance scores that are calculated. The center queries of each cluster are selected and the distance scores between each center query and a query Q is obtained to form a feature vector of Q, where each score is regarded as an attribute of the feature of Q. Thus the number of clusters equals to the number of dimensions (i.e., the number of feature attributes) of the feature vector of Q.

2.2.2.2   Template-Based Feature Modelling

The cluster-based feature modelling requires distances calculation between all training queries. Moreover, the clustering process adds additional time consumption. To reduce the feature modelling time, we propose to replace the center queries used in cluster-based feature modelling with representative queries that are generated by benchmark templates. Specifically, we generate queries from 18 out of 25 valid templates in the DBPSB benchmark [99] (we excluded queries which do not return any results: Query 1, 2, 3, 10, 16, 21 and 23). We refer to these queries as *template queries*. By recording the distance scores between a query Q with the 18 template queries, we obtain a 18-dimension feature vector for Q. The computation is then drastically reduced from $O(n^2)$ in cluster-based feature modelling to $O(n)$, where n is the number of queries. Therefore, our approach is feasible to apply to large size of training set.

Moreover, we adopt three dimension reduction algorithms, namely Canonical Correlation Analysis (CCA) [100], Principal Component Analysis (PCA) [101] and Non-negative Matrix Factorization (NMF) [102] on the feature vectors. In machine learning, dimension reduction is the process of reducing the number of random variables to describe a large set of data while still describing the data with sufficient accuracy. It helps reducing the learning time on the feature vectors. CCA calculates the coefficient among all features and chooses the most uncorrelated features. PCA aims to find a linear transformation to project the original data to a lower-dimensional space which is as informative as possible to capture as much of the variance of the original data in an unsupervised manner. NMF finds approximate decomposition of original data matrix and thus reduce the dimension by storing the two decomposed lower dimensional matrices.

Figure 2.7 illustrates the process and difference of cluster-based feature modelling and template-based feature modelling. In Fig. 2.7a, the modelling is based on all training queries. $\{D_{i=1}^{(n-1)n/2}\}$ record distances between each training queries. $\{C_{i=1}^{n}\}$ denote clusters. $\{d_{i=1}^{n}\}$ are distances between query Q and center queries of clusters. In Fig. 2.7b, t_1 to t_{18} are template queries. d_1' to d_{18}' are distances between query Q and 18 template queries. f_1' to f_r' are features that are obtained after applying dimensional reduction algorithm (i.e., CCA, PCA or NMF), where $r < 18$.

2.2.3   Suggesting and Prefetching Similar Queries

After the feature vectors are obtained (Step 1), we train a suggestion (or prediction) model with the feature vectors of training queries and suggest similar queries to a new query (Step 2).

We adopt k-Nearest Neighbours (k-NN) [91] as prediction model. k-NN is a non-parametric classification and regression algorithm that predicts the performance of new data point based on its k-nearest training data points:

$$p_{new} = \frac{\sum_{i=1}^{k}(p_i}{k}, \tag{2.8}$$

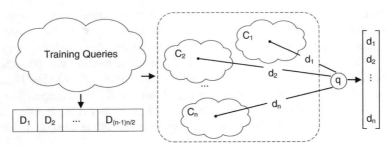

C_i : Cluster
d_i : Distance between q and center query of C_i
D_j : Distance between training queries

(a)

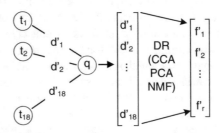

t_i : Template query
d'_i : Distance between q and t_i
f'_j : Feature after dimensional reduction on d'_1 to d'_{18}

(b)

Fig. 2.7 Feature modelling. The cluster-based modelling (**a**) is based on the distances among each pair of all the training queries. Clustering is done using these distances. d_i is the distance between query Q and the center query of the cluster C_i, and it represents a feature of Q. The template-based modelling (**b**) uses template queries. The distance between Q and template query t_i represents a feature of Q. After dimensional reduction (DR), the features are extracted to f'_1 to f'_r where $r < 18$

where p_{new} is the predicted value of the new data and p_i is the performance of the i-th nearest training data. If the new data is not in the training set, k-NN finds the point which is the closest to the new point according to its features.

k-NN is often successful in the cases where the decision boundary is irregular, which applies to SPARQL queries [90]. As it is originally a classification algorithm, one of the supervised learning algorithms, that requires labels or performance metrics of training data which we don't have, we therefore modify k-NN to build the k-d tree according to the Euclidean distance between feature vectors of SPARQL queries. The training thus turns into an unsupervised learning algorithm. Then we use trained k-NN model to suggest the nearest (i.e., the most similar) queries for a new query. Given the similar queries for a query Q, we prefetch the results of these queries directly from SPARQL endpoints and put the (q_i, r_i) pairs into the cache during the caching process, which is discussed in next section.

2.2.4   Caching and Replacement

As the cache has limit space, the less useful data should be removed from the cache to give space to more useful data. In traditional page-based cache in databases, the less useful data is less frequently accessed which will be removed to give space to more frequently accessed data. When the data required by a new query is in cache, the result is returned immediately. Each time the data in cache is accessed, it is called a cache hit. Therefore, the problem of cache replacement is the problem of identifying the more frequently accessed data.

SECF, the proposed client-side caching framework, does not require the knowledge of underlying system of SPARQL endpoint. Thus we cannot directly apply the cache replacement algorithms used in page-based databases. We propose to use a time-aware frequency based algorithm, which leverages the idea of a novel approach recently proposed for caching in main memory databases in *Online Transaction Processing (OLTP)* systems [95]. Specifically, we cache the (query, result) pairs and consider the hit frequencies of them when performing cache replacement. The recently most hit queries in the cache are *hot* queries which are more useful queries. Hot queries will be kept in the cache, whereas queries in the cache that do not belong to hot queries are considered less useful, which will be removed from the cache.

Before we introduce the proposed cache replacement algorithm in next section, we introduce how to measure the hit frequencies of queries first. Then we apply the proposed frequency measurement in developing two cache replacement strategies.

2.2.4.1   MSES

We adapt the algorithm used for identifying hot triples in our previous work [84]. Here we introduce this algorithm and how we adapt it to identifying hot queries.

The *Exponential Smoothing* (ES) is a technique to produce a smoothed data presentation, or to make forecasts for time series data, i.e., a sequence of observations [103]. It can be applied to any discrete set of repeated measurement and is currently widely used in smoothing or forecasting economic data in the financial markets. Equation (2.9) shows the simplest form of exponential smoothing. This equation is also regarded as *Simple Exponential Smoothing* (SES).

$$X_t = \alpha * x_t + (1 - \alpha) * X_{t-1}, \qquad (2.9)$$

where X_t stands for smoothed observation of time t, x_t is the actual observation value at time t, and α is a smoothing constant with $\alpha \in (0, 1)$. From this equation, it is easy to observe that SES assigns exponentially decreasing weights as the observation becomes older, which meets the requirement of selecting the most frequently and recently issued queries. The reason behind our choice of SES is its simplicity and effectiveness [95].

In SECF, we exploit SES to estimate hit frequencies of queries. In this case, x_t represents whether the query is hit at time t, thus it is either 1 for a cache hit; or 0 otherwise. Therefore, we can modify the Eq. (2.9) to *Modified Simple Exponential Smoothing* (MSES):

$$E_t = \alpha + E_{t_{prev}} * (1 - \alpha)^{t_{prev} - t} \tag{2.10}$$

where t_{prev} represents the time when the query is last hit and $E_{t_{prev}}$ denotes the previous frequency estimation for the query at t_{prev}. E_t denotes the new frequency estimation for the query at t [95].

The accuracy of MSES can be measured by its standard error.

We give the derivation of the standard error according to Eq. (2.10) and provide a theoretical proof that MSES achieves better hit rates than the most used cache replacement algorithm LRU-2 as follows.

The variance of the estimation E_t [104] is:

$$Var(E_t) = \sigma^2 \alpha^2 \left(\sum_{i=0}^{t-1} (1 - 2\alpha)^{2i} \right) \tag{2.11}$$

As

$$\sum_{i=0}^{t-1} (1 - 2\alpha)^{2i} = \frac{1 - (1 - \alpha)^{2t}}{1 - (1 - \alpha)^2} \tag{2.12}$$

we can get

$$Var(E_t) = \sigma^2 \frac{\alpha}{2 - \alpha} [1 - (1 - \alpha)^{2t}] \tag{2.13}$$

When $t \to \infty$, we get

$$Var(E_t) = \sigma^2 \frac{\alpha}{2 - \alpha} \tag{2.14}$$

Because the observation value can only be 0 or 1, it follows a *Bernoulli Distribution* (or *Two-point Distribution*), and the frequency p is actually the probability of the observation being 1. So, the standard error for this Bernoulli distribution is: $\sigma = \sqrt{p(1 - p)}$ [105]. Thus we can get the standard error of the estimation E_t as:

$$SE(E_t) = \sqrt{Var(E_t)} = \sqrt{p(1 - p)} \sqrt{\frac{\alpha}{2 - \alpha}} = \sqrt{\frac{\alpha p(1 - p)}{(2 - \alpha)}} \tag{2.15}$$

We also give the measurement of standard error for the most commonly used caching algorithm LRU-2 here as we compare our approach with LRU-2 in Sect. 2.3.3. In a recent work [95], the authors present a probability model for evaluating the distribution of LRU-2 estimation and find that it follows a geometric distribution. Thus its standard error is:

$$SE(LRU-2) = \sqrt{\frac{1-p}{p^2}} \tag{2.16}$$

By comparing Eqs. (2.15) and (2.16), it is easy to observe that $SE(LRU-2)$ is always bigger than $SE(E_t)$.

2.2.4.2 Cache Replacement Algorithms

We perform cache replacement based on the estimation score calculated by MSES. Each time a new query is executed, we examine the frequency of cache hit of this query using MSES. If it is in the cache, we update the estimation for it. Otherwise, we just record the new estimation. To decide which records can be kept in the cache, we develop two cache replacement strategies, namely the Full-records replacement and the Improved replacement. Both strategies use forward searching. There are three main advantages of the forward searching. Firstly, it is simple as we only need to choose a starting time and then calculate the new estimation when a query is hit again. Secondly, the forward searching enables us to update the estimation and the cache immediately after a new query is executed based on the previously recorded estimation. Thirdly, the forward searching implements an incremental approach that helps identify the warm-up stage and the warmed stage of the cache.

Full-Records Replacement In the full-records replacement, the algorithm keeps the estimation records for all processed queries. Specifically, it processes all the historical queries. When encountering a hit to a query at time t, the algorithm updates this query's hit frequency estimation using Eq. (2.10). When the scan is completed, the algorithm ranks each query by its estimated frequency and returns the H queries with the highest estimates as the hot set. These top-H queries are kept in the cache, while lower ranked queries will be removed from the cache. However, this algorithm requires storing the whole estimation record which is a large overhead. Furthermore, it consumes a significant amount of time when calculating and comparing the estimation values. To solve these issues, we consider improving the algorithm in two ways. One possible solution is that we just keep a record after skipping certain ones. This is a naive sampling approach. We vary the sampling rate but it turns out that the performance of this sampling approach is not desirable (see Sect. 2.3.3). The other possible approach is that we maintain partial records by only keeping those within a specified range of time which is improved caching replacement.

Improved Replacement In the improved replacement, we only keep estimation records from $t_{earliest}$ to the access time of the current processing query. The algorithm gets partial estimation records that are within the time range between $t_{earliest}$ and the hit time of the last processed query in the log. If the new estimated query is in the cache, it shows a cache hit, and the algorithm updates the new estimation calculated by Eq. (2.10) and the *last_hit_time* of this triple in estimation records. It calculates the new $t_{earliest}$ and t_{latest} if the new estimated triple holds the previous $t_{earliest}$ and t_{latest}. If $t_{earliest}$ is changed, estimation records with *last hit time* earlier than $t_{earliest}$ will be removed. If the new estimated query is not in the cache, which is a cache miss, the algorithm checks whether the new estimated query is in the estimation records. If so, it updates its estimation and *last_hit_time* in records. In addition, the cache needs to be updated if the estimation of the new estimated query is in the range of (est_{min}, est_{max}). This means it becomes a new hot query that should be placed into the cache. When the cache is updated, new $t_{earliest}$ and t_{latest} will be calculated, and the estimation records outside the time range will be removed. If the new estimated query is not in the estimation records, it needs to be added to the records.

2.3   Experimental Evaluations and Discussions

This section is devoted to the validation and performance study of our proposed approach. We first describe the setup of our evaluation environment in Sect. 2.3.1. In Sect. 2.3.2, we provide a detailed analysis of the real-world queries used in evaluation. Finally, we report the experimental results, including the performance comparison of cache replacement algorithms, feature modelling approaches and the comparison to the state-of-the-art work from Sects. 2.3.3 to 2.3.5.

2.3.1   Setup

Datasets We used real world queries gathered from USEWOD 2014 challenge. We analyzed the query logs from DBPedia's SPARQL endpoint[4] (DBpedia3.9) and Linked Geo Data's endpoint[5] (LinkedGeoData). The log files from these two datasets have the same format with four parts: anonymized IP address, time stamp, query, user ID. To extract queries, we processed the original query by decoding, extracting interesting values (IP, date, query string), identifying SPARQL queries from query strings and removing invalid queries (i.e., incomplete queries and queries with syntax errors according to SPARQL1.1 specification). We focused on SELECT queries in the experiments and retrieved 198,235 valid queries from DBpedia3.9 and 1,790,047 valid queries from LinkedGeoData. Within the SELECT

---

[4]http://dbpedia.org/sparql/.
[5]http://linkedgeodata.org/sparql.

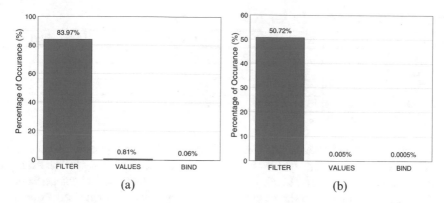

Fig. 2.8 Selected patterns from SELECT queries. FILTER occupies large proportion in SELECT queries. (**a**) DBPedia3.9. (**b**) LinkedGeoData

queries, except for patterns which can be finally decomposed to BGPs (e.g., AND, UNION, OPTIONAL and MINUS), FILTER, VALUES and BIND are used, especially for FILTER, which occurs in 83.97% SELECT queries in DBpedia3.9 query logs and 50.72% SELECT queries in LinkedGeoData (Fig. 2.8). This actually provides a strong evidence that FILTER expressions should not be ignored when calculating similarity between queries.

Implementation We obtained BGPs by parsing the SPARQL queries using Apache Jena-2.11.2. We implemented GED using a suboptimal solution integrated in the Graph Matching Toolkit.[6] The modified k-NN and LRU-2 were implemented in Java.

The System We set up our own SPARQL Endpoint by installing local Virtuoso server and loading datasets into the Virtuoso. The server has the configuration of 64-bit Ubuntu 14.4 with 16 GB RAM and 2.40 GHz Intel Xeon E5-2630L v2 CPU. Our code runs on a PC with 64-bit Windows 7, 8 GB RAM and 2.40 GHZ Intel i7-3630QM CPU.

2.3.2 Analysis of Real-World SPARQL Queries

2.3.2.1 Analysis of Average Queries

We used the distance measurement described in Sect. 2.2.1 to cluster the queries. Table 2.1 shows that the average queries for a client in DBpedia3.9 log files is 23.31, and the average clusters a client's queries belong to is 2.27. The average queries in

---

[6]http://www.fhnw.ch/wirtschaft/iwi/gmt.

Table 2.1 Analysis of clients associated with queries and clusters

	AvgQ/Client	AvgClusters/Client	AvgQ/Cluster
DBpedia3.9	23.31	2.27	10.26
DBpedia3.9-100	447.49	3.23	138.54
LinkedGeo	612.51	5.16	118.70
LinkedGeo-100	16,441.84	67.90	245.40

each cluster is 10.26. For LinkedGeoData queries, the average queries per client is 612.51 and each client's queries belong to 5.16 clusters in average. The average queries in each cluster is 118.70. Our analysis shows that each client performed several queries which have shared clusters, indicating clients issued similar queries.

We also selected 100 clients with the most queries issued and found that the average number of queries for these top clients in DBpedia3.9 is 447.49 and the average clusters for one client is 3.23. The average queries per cluster increases to 118.70. The numbers for the top 100 clients who issued most queries in LinkedGeoData are higher, as average queries per client is 16,441.84, and average number of clusters per client is 67.90. In average, 245.40 queries belong to one cluster. These findings indicate that clients performed a large number of queries which have similar structures. The average number of queries for one cluster shows that queries have high similarity.

2.3.2.2   Analysis of Subsequent Queries

To estimate how likely it is that queries are similar to previous queries and therefore could benefit from the prefetched results, we evaluated the time difference between one query and the next matched query (i.e., the next query belonging to the same cluster) for each client. DBpedia3.9 log files have a total of 8500 distinct client IDs and LinkedGeoData has 2921 distinct client IDs. We assigned distinct IDs starting from 1 for clients of both datasets. Figure 2.9 shows the average time gap between two matches for both datasets. Blue crosses indicate the average time gap (to next matched query) for each client ID. From Fig. 2.8a, we can see most of the time gap are close to zero, positioning blue cross on the X-axis, which means that the client issues similar-structured queries subsequently. A small number of blue crosses are away from the X-axis, which means that these clients seldom issue similar-structured queries. Similar observations are found in Fig. 2.8b. It demonstrates the fact that most clients issue similar queries subsequently. Moreover, on average, the similar queries are issued by the same client with previous queries within a very close time period. This further confirms that our approach can benefit the clients' subsequent queries in terms of querying answering speed.

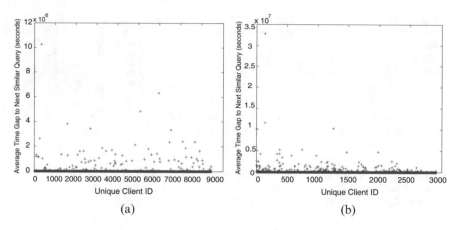

Fig. 2.9 Analysis of average matched time. Similar queries are usually issued within a short period of time. (**a**) DBPedia3.9. (**b**) LinkedGeoData

2.3.3 Performance of Cache Replacement Algorithm

We firstly evaluated our cache replacement algorithm MSES, because it would be used in the following experiments. To evaluate the performance of MSES, we implemented various algorithms including the full-record MSES, improved MSES, the sampling MSES, and LRU-2. LRU-2 is a commonly used page replacement algorithm which we implemented based on record rather than page. All of Linked-GeoData valid queries obtained were processed in this experiment because the size of this query set is much larger than DBpedia3.9 query set. Thus we can observe the difference between Improved MSES and MSES algorithms.

2.3.3.1 Impact of α

As the *Exponential Smoothing* has only one parameter α, the choice for α would affect the hit rate performance. However, as per our experiments on different values for α, the hit rates differ only slightly and a value of 0.05 shows better performance, as shown in Fig. 2.10a.

2.3.3.2 Impact of Cache Replacement Algorithms

Figure 2.10b shows the hit rates achieved by the four algorithms we implemented. It should be noted that in the experiment, the caching size was set to 20% of the total historical queries and α was set to 0.05 for MSES and its variants. We chose 20% as the caching size because it is neither too large (e.g., >50%) to narrow the performance differences among algorithms, nor too small (e.g., <10%) leading

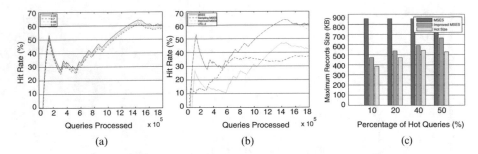

Fig. 2.10 Cache replacement performance (LinkedGeoData). Different α have different impact on the hit rates (**a**). Different cache replacement algorithms affect the hit rates largely (**b**). The improved MSES reduces the space overhead largely compared to the MSES (**c**)

to inaccurate performance evaluation due to insufficient processed data. From the figure, we can see that MSES and Improved MSES have the same hit rate until they have processed about 1.4 million RDF triples, after which MSES has a higher hit rate than Improved MSES. This is because MSES maintains the estimations for all processed records while the Improved MSES only keeps part of the estimations. The changing point denotes that from which, the Improved MSES maintains partial volume of estimation records. From the figure, we can also see that Sampling MSES does not perform well. This figure only shows the hit rate of sampling MSES with the sampling rate of 50%, which is expected to have a high hit rate. The LRU-2 algorithm has the lowest hit rate of all the algorithms. The hit rates of all algorithms start from 0 and reach their first peak at certain points, then fluctuate. The direction to the first peak shows the warm-up stage and the rest of the lines are the warmed stage. This illustrates that we exploit an incremental approach, which includes a warm-up stage to calculate the hit rate.

2.3.3.3 Space Usage for Records

Figure 2.10c gives the measurement of space usage by recording the estimations. As discussed before, MSES performs better than the Improved MSES. However, it consumes more storage space to maintain the estimation records for all processed triples. It also takes longer time to check the cache. Figure 2.10c shows the maximum space consumption for each algorithm. Note that we used all valid LinkedGeoData queries in this experiment. The columns are classified into four groups which represent the percentage of hot queries to all processed queries. In each group, the left column represents the maximum space used by MSES, including the hot queries and the estimation records. The middle column represents the space usage of the Improved MSES that also includes the hot queries and the estimation records. The right column represents the size of the hot queries. From this figure, we can see that the Improved MSES consumes less space.

2.3.4 Comparison of Feature Modelling Approaches

In the experiments of this section, we compared our feature modelling approach (i.e., template-based feature modelling) with the state-of-the-art approach (i.e., cluster-based feature modelling), and evaluated the performance under the scenarios of applying and without applying suggestion/prefetching. We applied the dimensional reduction algorithms on both template-based feature modelling and cluster-based feature modelling.

Because the time consumption of cluster-based approach is tremendous, we did not use all valid queries as the training set. We randomly chose 21,600 training queries and 5400 testing queries from the two query sets separately. The cache replacement algorithm we used in all testing cases is Improved MSES and we chose $\alpha = 0.05$. Because the larger size of cache, the higher hit rate would achieve, we only show experiment results when the number of queries in cache is set to 1000.

2.3.4.1 Performance of Cluster-Based Feature Modelling

Impact of Cluster Number As the generation of graph feature vectors depends on the number of clusters, we compared the impact of different number of clusters on the average hit rates. Figure 2.11 shows the hit rates with different number of clusters (5, 10, 15, 20, 30 in our experiments) and different k in k-NN (we chose 2, 5, 10, 20, 50, 100 as k). Although the hit rates on DBpedia3.9 queries change slightly when using the same k with different clusters, we can find that Cluster 10 (C10 for short thereafter) gives the highest performance from 77.18% ($k = 2$) to 94.18% ($k = 100$). For LinkedGeoData queries, the highest hit rate is achieved by C15 from 3.80% ($k = 2$) to 9.40% ($k = 100$). Note that the hit rate increases when the value of k increases for both DBpedia3.9 and LinkedGeoData queries. In the following experiments, we used C10 for DBpedia3.9 query and C15 for LinkedGeoData for feature vector generation. Moreover, Fig. 2.11 shows that the hit rates performed without suggestion is always lower than the ones with suggestions because cached similarly-structured queries contribute to the improvement of hit rates.

Impact of Dimension Reduction In order to compare to template-based feature modelling approach, we also applied dimensional reduction algorithms on cluster-based feature modelling approach. We generated new feature files with different lower dimensions for DBpedia3.9 and LinkedGeoData queries using CCA, PCA and NMF discussed in Sect. 2.2.2.2. The files are from Dimension 1 (D1) to D9 for DBpedia3.9 with 10 clusters (C10) and D1–D14 for LinkedGeoData with 15 clusters (C15). We then trained k-NN model with these files respectively and got k suggested queries for a randomly chosen query Q. We computed the average distance between suggested queries with Q and computed the distances obtained when using CCA, PCA and NMF. The lower the average distance is, the better the suggestion is. As large amount of the queries from these two SPARQL endpoints are similarly-structured or repeated (see Sect. 2.3.2), we set a large number of queries

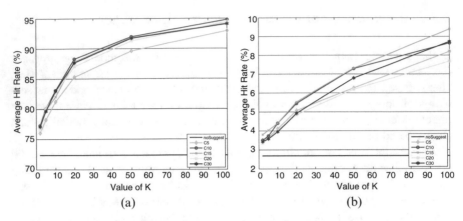

Fig. 2.11 Hit rates with different clusters and k. (**a**) DBPedia3.9. (**b**) LinkedGeoData

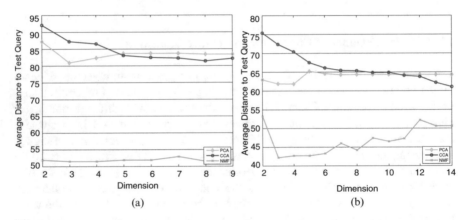

Fig. 2.12 Performance comparison among using CCA, PCA and NMF to reduce dimension (cluster-based). (**a**) Distances for C10 (DBpedia3.9). (**b**) Distances for C15 (LinkedGeoData)

to avoid the distance to be zero. Thus we chose $k = 500$ in k-NN in this experiment. As shown in Fig. 2.12, NMF always performs the best for both DBpedia3.9 and LinkedGeoData queries. It gets optimal result when the number of dimensions is 3. PCA performs better than CCA when dimension is low and worse than CCA when dimension becomes high. For DBpedia3.9 queries, the intersection is $D = 5$, while for LinkedGeoData, the intersection is $D = 11$. We used NMF for our dimension reduction in the comparisons thereafter.

2.3.4.2 Performance of Template-Based Feature Modelling

In template-based feature modelling, we also leveraged dimensional reduction algorithms. The performance of different algorithms is shown in Fig. 2.13. It

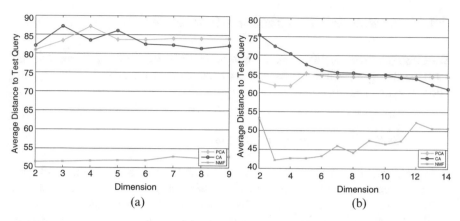

Fig. 2.13 Performance comparison among using CCA, PCA and NMF to reduce dimension (template-based). (**a**) Distances (DBpedia3.9). (**b**) Distances (LinkedGeoData)

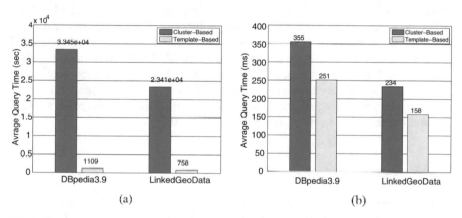

Fig. 2.14 Time comparison on feature modelling approaches. (**a**) Training time on 21,600 training queries. (**b**) Average query time

is shown that NMF still outperforms other algorithms in extracting the most representative features.

Figure 2.14 gives the impact of two feature modelling algorithm on time consumption. Figure 2.14a compares training time on 21,600 training queries. Cluster-based approach requires 33,446 s, which is more than 9 h for DBPedia3.9 queries, and 23,405 s (i.e., more than 6 h) for LinkedGeoData queries. Template-based approach largely reduces the time to 1109 and 758 s, respectively. Figure 2.14b evaluates the average query time. Our template-based approach also outperforms the cluster-based approach.

2.3.5 Performance Comparison with the State-of-the-Art

We also compared our work with the Adaptive SPARQL Query Cache (ASQC) introduced in [14], as it is the first and complete work to cache SPARQL query in a client-side manner.

2.3.5.1 System Performance Comparison

In this experiment, we compared the average hit rate, average query time and space usage between our work SECF and ASQC. We also gave measurement when no cache was used. To compare our approach with ASQC, we modified the code of ASQC[7] to access our datasets. We performed the experiment on DBpedia3.9 dataset. We used Cluster-Based Feature Modeling, and Improved MSES with $\alpha = 0.05$. Table 2.2 presents the results. ASQC has slightly lower hit rate (72.63%) than SECF (78.59%). ASQC takes 264 ms in average for one query and SECF takes 247 ms. When no cache is implemented, the average query time increases to 625 ms. We did not include prefetching time as it is in a separate thread. Space consumption evaluates how much memory the cache uses. In SECF, the total usage (before slash) for caching 1000 queries, as shown in Table 2.2, includes cached queries and answers as well as the estimation records for cache replacement (after slash). We used the same implementation for cache in order to compare. The results indicate that most space is consumed by cached (query, result) pairs.

2.3.5.2 Server Overhead Comparison

In order to evaluate the impact of cache on the endpoint server, we monitored the memory and CPU usage as well as I/O on the server. We captured the usage every 20 s until the querying ends. Table 2.3 shows the average free memory (AvgFreeMem), average I/O (AvgFreeMem) and average CPU time (AvgCPU) including system CPU and user CPU time. We only present the result on querying DBpedia3.9 dataset due to limit space. From the result we find out that SECF and ASQC cause higher computation overhead (I/O and CPU) and memory usage on server compared to querying without cache and ASQC performs slightly better

Table 2.2 Performance comparison

	ASQC	No cache	SECF
Hit	72.63%	NA	78.59%
AvgTime	264 ms	625 ms	247 ms
Space	7.15 MB	NA	7.15 MB/0.45 KB

---

[7]http://wiki.aksw.org/Projects/QueryCache.

Table 2.3 Server
performance comparison

	ASQC	No cache	SECF
AvgFreeMem	217.87 MB	224.30 MB	206.35 MB
AvgIO	11.49	7.72	21.43
AvgCPU	9.37 ms	10.09 ms	10.60 ms

than SECF with more free memory (217.87 MB vs 206.35 MB), less I/O (11.49 vs 21.43) and less CPU time (9.37 ms vs 10.60 ms). It is because that SECF requires prefetching results for similar queries from server which leads to additional overhead.

2.3.6   Discussions

In this section, we discuss some issues from our experience in this work.

Partial Caching vs Query Caching   Some works focus on caching part of queries (i.e., subgraph) [16, 17] and identifying the hit subgraph (i.e., containment checking) [15]. These methods not only cache the exactly matched queries, but also the queries which share the same subgraph with the cached ones. Therefore, using such subgraph caching broadens the cached items and improves the cache hit rate. However, the parsing and identifying of query subgraphs is a very time-consuming task that counteracts the speed improvement achieved by cache. In some occasions it cannot accelerate the querying speed. In our current work, we have not considered partial caching of queries. Investigating efficient solutions by integrating partial queries into our framework will be part of our future work.

Dynamic Learning   In the training process, the larger the size of the training queries, the better performance we can get. The reason is that more query variety can be captured and the model will be less sensitive to unforeseen queries. However, in practice, unseen queries can be issued very frequently and quickly. In this case, no matter how large the current training set is, it is inefficient to cover as much as possible query variety. There are two solutions for this issue. One is periodically training on historical queries. We adopt this idea in our work. Similar to the suggestion process which is in a background thread, the periodical training, especially the building of the k-NN model, also runs in the background. Therefore, although it is time-consuming to train on large query sets, it will not affect the querying process. One benefit to use periodical training is that we can leverage the well developed learning algorithms. Moreover, our approach have already achieved great improvements in reducing the training time (Sect. 2.3.4). The other solution is incremental learning (or referred to online learning), which holds a new input batch in addition to the existing learning model to reflect recently executed queries. We leave the investigation of these techniques (e.g., [106]) as our future work. Regarding the dynamics of the server data, it does not affect the performance of our approach. Because dynamic server data affects the performance of the query execution which is not considered in this work.

Space Overhead From evaluation (Sect. 2.3.5.2), we observe that the memory space used by SECF is mostly consumed by cached (query, result) pairs. This is because SECF caches the pairs in text directly. The size of the cached text can be reduced by leveraging encoding techniques. Many existing triple stores (i.e., systems that holds RDF data) [63, 71] encode the RDF triples to numerical values in order to reduce the space overhead. By developing their own indexing algorithms, the access and retrieval of the triples become efficient. We could adapt the ideas to encode SPARQL queries or part of queries, e.g., BGPs and develop tailored indexing algorithm. We will investigate these techniques in the future.

Sever Overhead We observe from the comparison to the state-of-the-art work, ASQC, that SECF introduces larger overhead to server side (Sect. 2.3.5.2). This is due to the fact that the prefetch process continuous requests data directly from the SPARQL endpoint server. One possible improvement for this issue is to find the common results of multiple similar queries, so that less data requests will be issued. Containment checking can be considered to solve this problem [15], however, it is time-consuming. A supplementary solution is to add estimation of the results of these queries and prune the ones that would return empty results. In this way, less queries are issued directly to the server side. We will further investigate these technique and leave it as our future work.

2.4 Related Work

Our work mainly addresses caching problems in two research areas, namely *Semantic Caching* and *Query Suggestion*. In this section, we review the recent representative works that are related to our work.

2.4.1 Semantic Caching

Semantic Caching involves techniques that keep previously fetched data for past queries. If subsequent queries use the same data, results can be returned immediately. It was originally developed for relational databases. Godfrey and Gryz [107] present a predicate-based caching schema in client server, which is a general logical foundation for semantic query caching. It is a comprehensive framework that addresses multiple issues regarding build and recover cache and the notions of semantic overlap and independence. Dar et al. [108] present a semantic schema of caching for client-server systems. They cache semantic regions rather than tuples or pages. They provide distance-based cache replacement policy, in which the distance that is farthest from the client's current location is discarded from cache. This approach is designed for SELECT queries.

In recent years, the semantic caching technique has been extended to triple stores that manage SPARQL queries. The work of Martin et al. [14] is the first step towards semantic caching for SPARQL queries, in which both the complete triple query result and the application object are cached. The work essentially builds a proxy layer between an application and a SPARQL endpoint. It is a client-side caching method where we draw our idea. The cache layer considers only the identical queries issued afterwards and identical application object that could be potentially used. Shu et al. [15] improve this approach in a content-aware way by introducing query containment checking which evaluates whether a query can be answered by the result of a cached query. Thus not only the identical queries, but also the queries which succeed in the containment checking can utilize the result of the cached queries. But this method introduces large overhead on containment checking itself. Yang and Wu [16] develop an approach that caches intermediate result of basic graph patterns in SPARQL queries. It decomposes the query into BGPs and evaluates if the result of any BGP or join of BGPs is cached. The cached results which are hit will be returned and joined with the other parts of the query to form the final query result. This work does not address the impact of different join orders. It is a server-side caching method that embeds the proposed cache in the RDF engines. Very recently, Papailiou et al. [17] introduce canonical labelling to identify isomorphic subgraphs in SPARQL query patterns, which are cached for subsequent querying. This solution focuses on implementing a caching layer on top of the distributed partitions and dynamically updating the contents of the cache. It is a server-side caching method that is designed to be embedded in the distributed RDF engines.

The Linked Data Fragments (LDF) approach [88] aiming at improving data availability can also be regarded as caching technique as it caches fragments of queryable data from servers that can be accessed by clients. In this way, each client is able to process SPARQL queries on replicated fragments cached from servers. This approach can be potentially used for federated queries. However, performance degradation issue needs to be addressed.

2.4.2   Query Suggestion

Query suggestion is an interactive approach used in search engines to better understand users' information needs. It plays an important role in improving the accuracy of searching. Query suggestions are usually made by mining query logs and session records of Web users' searching history [109]. They either aim to find similar queries in search logs and use those queries as suggestions, or identify pairs of queries which co-occur in the same query sessions.

Researchers recently have introduced these mining techniques into SPARQL processing. Lehmann and Bühmann [51] propose a novel solution for querying knowledge bases. It leverages supervised machine learning to suggest SPARQL queries based on examples previously selected by users. This approach narrows the range of possible answers asked by users. With the learning techniques, no prior knowledge of the underlying schema or the SPARQL query language is required. More recently, Hasan [90] uses a machine learning method to predict the performance of SPARQL query performance. Specifically, a suggestion model is trained with previously issued queries and then based on this, the query time for new queries can be predicted. In our work, we extend this approach to suggest similarly-structured queries and prefetch and cache their answers.

Query relaxation (also called query expansion) is closely related to query suggestion, as it extends the original query with related information, so that it accelerates the overall query processing. Its aim is to improve the recall of query. Features with similar meaning need to be identified and suggested to generate expanded query. In recent years, query expansion techniques have been used by several research efforts that focus on topics of SPARQL queries. Elbassuoni et al. [111] propose multiple types of relaxation to improve the recall of entity-relationship search. Lorey and Naumann [87] cluster similar SPARQL queries to different templates in order to detect recurring patterns in queries. These templates can be used to expand queries for query processing. Fokou et al. [110] investigate query relaxation over RDF data and focus on identifying parts of SPARQL query that are responsible of the failure of the query. It aims at providing users with alternative answers instead of an empty result.

2.5 Summary

In this chapter, we have utilized machine learning techniques to learn clients' query patterns, which can be used in identifying potential subsequent queries. These queries are then issued in a background thread and their results are prefetched and cached in order to reduce the overall querying time. Our proposed template-based feature modelling method greatly outperforms the state-of-the-art method, cluster-based feature modelling method in terms of training and suggestion time. We have also proposed a cache replacement algorithm, MSES, tailored for the our cache content. MSES is simple yet effective. The comparison to the state-of-the-art client-side caching framework ASQC shows that SECF outperforms ASQC in terms of the average query time, but requires more system overhead on the server which we will tackle in future works.

We have discussed how we improved the query performance on curated KBs in this chapter. In the next chapter, we will discuss our method to predict the query performance, which helps pre-arrange the resources effectively.

Chapter 3
Query Performance Prediction on Knowledge Base

In this chapter, we adopt machine learning techniques to predict the performance of SPARQL queries. Our work focuses on modelling features of a SPARQL query to a vector representation and use these feature vectors to train predictive models. This method does not depend on underlying systems and any knowledge of the underlying data, but only on the nature of SPARQL queries. We adopt multiple regression models as prediction models and propose a one-step and a two-step prediction processes. Query performances in both cold and warm stages are studied. Evaluations are performed on real world SPARQL queries, whose execution time ranges from milliseconds to hours. The results demonstrate that the proposed approach can effectively predict SPARQL query performance and outperforms state-of-the-art approaches. This chapter is based on our research reported in [112, 113].

3.1 Design Overview

The Semantic Web, with its underlying data model Resource Description Framework (RDF) and its query language SPARQL Protocol and RDF Query Language (SPARQL), has received increasing attention among researchers and data consumers in both academia and industry. Over the recent years, RDF has been increasingly used as a general data model for conceptual description and information modelling. For example, knowledge management applications such as DBpedia[1] and Freebase[2] offer large collections of facts about entities and their relations with RDF-based representations. Domain knowledge bases provide biology resources (e.g., UniProt,[3]

---

[1] http://dbpedia.org/.

[2] https://www.freebase.com/.

[3] http://www.uniprot.org/.

© Springer International Publishing AG, part of Springer Nature 2018
W. E. Zhang, Q. Z. Sheng, *Managing Data From Knowledge Bases: Querying and Extraction*, https://doi.org/10.1007/978-3-319-94935-2_3

BioPortal[4]) and spatial data (e.g., LinkedGeoData[5]). Since the number of publicly available RDF datasets and their volume grow dramatically, it becomes essential for efficient querying of large scale RDF datasets. This is an important issue in the sense that whether we can obtain knowledge efficiently affects the adoption of RDF data as well as the underlying Semantic Web technologies.

3.1.1   Motivation

Substantial works focus on the prediction of query performance (e.g., execution time) [114–117]. Prediction of query execution performance can benefit many system management decisions, including:

- Workload Management: Predicting the execution performance accurately before executing incoming queries can help estimate workloads and effectively arrange available resources.
- Query Scheduling: Understanding the query performance metric of an incoming query can help decide whether and when to run the query to avoid system hanging. The long-running query can be rewritten in order to improve performance.
- System Sizing: The sizing of systems (e.g., CPU, Memory etc.) is dependent on the peak value of resources required to complete unforeseen queries.
- Capacity Planning: Given an expected change to a workload, the decision on whether to upgrade the system for required resources depends on the accurate estimation of query execution performance.

Studies show that cost model based query optimizers are insufficient for query performance prediction [118, 119]. Therefore, approaches that exploit machine learning techniques to build predictive models have been proposed [118–120]. These approaches treat the database system as a black box and focus on learning query performance prediction models, which are evaluated as feasible and effective [118]. These works extract the features of queries by exploring the query plan that can provide estimated values such as estimated execution time, estimated row count and these two estimations for each query operator (e.g., AND).

For SPARQL queries, the query engines can be grouped into two categories: RDBMS-based and RDF native triple stores. RDBMS-based engines rely on optimization techniques provided by relational databases. However, due to the absence of schematic structure in RDF, cost-based approaches show problematic query estimation which cannot effectively predict the query performance [121]. RDF native query engines typically use heuristics and statistics about the data for selecting efficient query execution plans [122]. Heuristics and statistics based

---

[4]http://bioportal.bioontology.org/.

[5]http://linkedgeodata.org/.

optimization techniques generally work without any knowledge of the underlying data, but in many cases, statistics are often missing [121].

Hassan [90] proposes the first work on predicting SPARQL query execution time by utilizing machine learning techniques. In the work, multiple regression using Support Vector Regression (SVR) is adopted. The evaluation is performed on benchmark queries on an open source triple store Jena TDB.[6] The feature models are extracted based on *Graph Edit Distances* (GED) between each of training queries. However, in practice, we observe that the calculation of GED is very time consuming, which is not a desirable method when the training dataset is large. Moreover, their work omits the study of the cold stage of the system, where query compilation time should not be ignored. In this work, we will investigate the total elapsed time that includes both compilation time and execution time of a query. When the system is in the warm stage of query processing, i.e., the query is not executed for the first time, the compilation time is omitted, thus the elapsed time is equal to the execution time. When the query is new to the system, i.e., in the cold stage of the system, the compilation time cannot be ignored when examining total elapsed time. Although we discuss elapsed time in our work, we will use execution time hereafter for expressivity. In addition to execution time, other performance metrics such as CPU usage and memory usage also need to be considered.

3.1.2 Challenges

To effectively predict SPARQL query performance, we draw ideas from Hassan but adopt different machine learning techniques to address the issues of training large datasets and investigation of the cold stage of the system. The challenges in our work center on capturing characteristics of SPARQL queries and representing the characteristics as features for the application of machine learning techniques.

- Feature Obtaining: Our aim is to promote higher usability of semantic web and more effective consumption of RDF-represented information, thus the work on open source triple stores is more applicable. The intuitive solution for obtaining feature is to leverage the query plan provided by triple stores. Similar works for SQL queries are presented in [118, 119]. However, explicit plan information is hardly available in open source triple stores (we have examined four most commonly used open source triple stores Virtuoso (open source edition),[7] Fuseki,[8] Sesame[9] and 4Store[10]). Furthermore, as aforementioned, if the plan

---

[6]https://jena.apache.org/documentation/tdb/.

[7]http://virtuoso.openlinksw.com/.

[8]http://jena.apache.org/documentation/serving_data/.

[9]http://rdf4j.org/.

[10]http://4store.org/.

information is explicit, it is still based on cost model estimations, which are shown to be ineffective. Therefore, we cannot leverage query plans to construct the feature of a SPARQL query. The question then is: What else information can we obtain from a SPARQL query that captures the characteristics of the query?

- Feature Representation: Given the feature acquired from a SPARQL query, a vector representation is required for machine learning algorithms. How can we convert the features to a feature vector that effectively represents the query without losing information?
- Feature Extraction: The training of prediction model is based on the query features. Irrelevant features will introduce noise in the training process, which leads to distortion of prediction results. How can we select the most predictive features?

3.1.3 Prediction Approach Overview

To address the above challenges, we propose three approaches for modelling features, namely query algebra features, Basic Graph Pattern (BGP) features and hybrid features. Specifically, we transform the algebra and BGPs into a feature vector. We propose a feature selection process based on heuristic to build hybrid features and also compare the feature selection and extraction approaches on the performance of prediction. Once the features are built, they can then be used to estimate the performance of a new requested query based on the feature values that can be obtained without executing the query. We consider both k-nearest-neighbour (k-NN) regression and SVR as prediction models. Both average k-NN and weighted k-NN are investigated.

To improve the prediction performance, we develop a two-step prediction process in addition to one-step prediction, depicted in Fig. 3.1. The figure shows that our prediction process consists of three main components, namely data pre-processing, feature modelling and prediction model training and prediction. Both training queries and new requested queries are cleaned in the data pre-processing component. In the training component, prediction models are derived from the training queries with observed query performance metrics. In this component, queries are represented as a set of features (i.e., predictive variables) with corresponding performance metrics (i.e., target variables). A feature matrix is obtained with each row representing a feature vector of a query. The columns are instances of different features. After obtaining features, we further apply feature selection to reduce the dimension of the feature matrix to extract the most predictive features. In the one-step prediction, feature matrix of the training queries is fed into the prediction model training component, with the goal of creating a mapping between feature values and observed query performance metrics. The prediction models are then used to predict the performance of new requested queries. In the two-step prediction, a classification step is added before training the model. The aim of classification is to group queries with different ranges of execution time. Multiple models are trained for different

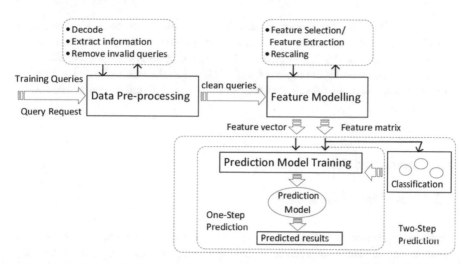

Fig. 3.1 SPARQL query performance prediction

classes and different performance metrics separately. In this work, we classify the queries into four classes and we perform prediction for execution time, CPU usage and memory usage of SPARQL queries.

For the detailed discussion of the three components, we focus on discussion of feature modelling and prediction model training in Sects. 3.3 and 3.4 and provide a brief introduction of data processing in experiment set-up.

Our approach can be applied in the situation that no estimations of query execution performance are provided, or such estimations are implicit or inaccurate. In practice, this applies to most triple stores that are publicly available. In a nutshell, the main contributions of this work are summarized as follows:

- We adopt machine learning techniques to effectively predict the query performance before their execution. Rather than only predicting query performance on the warm stage, we also consider the cold stage query execution, which is not discussed in the state-of-the-art works. The proposed methods are easy to reproduce as we mainly adapt the most commonly used algorithms in the machine learning field. Furthermore, little domain expertise is required.
- We propose three ways to model features of a SPARQL query. The algebra and BGP features can be acquired from the parsing of the query. The hybrid feature can be derived from algebra and BGP features. All features can be easily obtained without the information provided by the underlying systems.
- We perform extensive experiments on real queries obtained from widely accessed SPARQL endpoints. The triple store we used is one of the most used systems in the community of Semantic Web. Thus our work will benefit a large population of users. Our approach is transferable and can be applied to other triple stores.

The remainder of this chapter is structured as follows. In Sect. 3.2, we briefly introduce the basic techniques used in this work. Then we introduce our feature modelling approaches in detail in Sect. 3.3. Our prediction approaches are described in Sect. 3.4, followed by the report of experimental evaluation in Sect. 3.5. We also discuss some issues we observed in this work in Sect. 3.6 and overview the related works in Sect. 3.7.

3.2 Preliminaries

3.2.1 Multiple Regression

Multiple regression focuses on finding the relationship between a dependent variable and multiple independent variables (i.e., predictors). It estimates the expectation of the dependent variable given the predictors. Given a training set $(\mathbf{x}_i, y_i), i = 1, \ldots n$, where $\mathbf{x}_i \in \mathbb{R}^m$ is a m-dimensional feature vector (i.e., m predictors), the objective of multiple regression is to discover a function $y_i = f(\mathbf{x}_i)$ that best predicts the value of y_i associated with each $\mathbf{x}_i$ [123].

Support Vector Regression (SVR) is to find the best regression function by selecting the particular hyperplane that maximizes the margin, i.e., the distance between the hyperplane and the nearest point, called support vectors [124]. The error is defined to be zero when the difference between actual and predicted values are within a certain amount ξ. The problem is formulated as an optimization problem:

$$\min \mathbf{w}^T \mathbf{w}, \quad s.t. \quad y_i(\mathbf{w}^T \mathbf{x}_i + b) \geq 1 - \xi, \xi \geq 0 \tag{3.1}$$

where parameter $\frac{b}{\|\mathbf{w}\|}$ determines the offset of the hyperplane from the origin along the normal vector $\mathbf{w}$. If we extend the dot product of $\mathbf{x}_i \cdot \mathbf{x}_j$ to a different space of larger dimensions through a functional mapping $\Theta(\mathbf{x}_i)$, then SVR can be used in non-linear regression. $\Theta(\mathbf{x}_i) \cdot \Theta(\mathbf{x}_j)$ is called kernel function. An advantage of SVR is its insensitivity to outliers [123].

k-Nearest Neighbours (k-NN) is a non-parametric classification and regression method [91]. The k-NN regression predicts based on k nearest training data. It is often successful in the cases where the decision boundary is irregular, which applies to our training data [90]. By training the k-NN model, the predicted query time can be calculated by the average time of its k nearest neighbours:

$$t_Q = \frac{1}{k} \sum_{i=1}^{k} (t_i), \tag{3.2}$$

where t_i is the execution time of the ith nearest query.

3.2.2   Dimension Reduction

In machine learning, dimension reduction is the process of reducing the number of random variables to describe a large set of data while still describing the data with sufficient accuracy. Dimension reduction is often performed before other machine learning tasks, such as clustering, classification and prediction. Other benefits include enhancing the interpretability of data, reducing over-fitting and shortening the training times [125].

Dimension reduction is divided into two sub-types: feature selection and feature extraction. Feature selection returns a subset of the features and the feature selection techniques are often used in the domains where representative features need to be identified, such as weight and height of subjects in healthcare. Feature extraction creates new features from original features by transforming the original features in a high-dimensional space to a space of lower dimensions. The transformation may be linear or non-linear. It is often used to avoid the effects of the curse of dimensionality [123].

3.3   Feature Modelling for Queries

The prediction relies on the features in the training sets, thus the performance of prediction is highly dependent on how much information the features can obtain from the data and how well the features represent the data. In order to utilize machine learning algorithms for SPARQL query performance prediction, we transform the query into vector representation. We formulate the problem as follows:

Definition 3.1 (SPARQL Feature Modelling) Let $Q = (F, P)$ denote a SPARQL query, where F is the SPARQL query form in {Select, Describe, Construct and Ask}. P is the query pattern of Q, feature modelling is the transformation that maps $Q \rightarrow \mathbf{q}$, where $\mathbf{q} \in \mathbb{R}^m$ and m is the number of features.

In this study, we use only static, compiling time features which can be extracted prior to execution. Algebra features and BGP features are obtained by parsing the query text (see Sects. 3.3.1 and 3.3.2). Hybrid features are generated by applying the selection algorithm we develop based on algebra and BGP features (see Sect. 3.3.3).

3.3.1   Algebra Features

The step of parsing the query string to algebras is prior to optimization by query engines. The algebras can be presented as a tree:

Definition 3.2 (Algebra Tree) Given a SPARQL query Q, algebra tree $T_{Algebra}(Q)$ is a tree where the leaves are BGPs and nodes are algebra operators presented hierarchically. The parent of each node is the parent operator of current operator.

We obtain the SPARQL algebra tree and then traverse the tree to construct the *Algebra Set* by recording the occurrences and hierarchical information of each algebra operator.

Definition 3.3 (Algebra Set) Given an algebra tree denoted as $T_{Algebra}(Q)$, the Algebra Set is a set of tuples $\{(opt_i, c_i, maxh_i, minh_i)\}$, where opt_i is the operator name, c_i is the occurrence count of opt_i in $T_{Algebra}(Q)$, $maxh_i$ and $minh_i$ are opt_i's max height and min height in $T_{Algebra}(Q)$, respectively.

We then transform the algebra set to a vector by concatenating all the tuples in the algebra set sequentially. We further insert the tree's height at the beginning of the vector. The values of each position is considered as an instance of a feature, thus we obtain a feature vector for a query. Figure 3.2 illustrates an example of

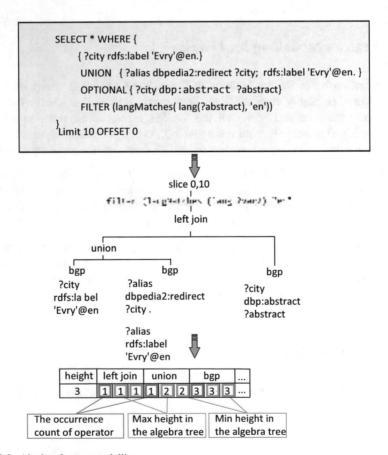

Fig. 3.2 Algebra feature modelling

algebra feature modelling. We have a 40-dimensional feature vector for the example SPARQL query.

3.3.2 BGP Features

Algebra features take occurrences and hierarchical information of operators into consideration. For complementary, we propose to leverage graph structure of BGPs, the most used subset of SPARQL queries [126] to form BGP features. We propose to capture the characteristics of BGPs and transform them into a vector representation. We firstly examine BGPs that consist of sets of triple patterns. We could follow the way in the algebra feature modelling to count the number of occurrences of triple patterns. However, in this way, we fail to represent the hierarchy of triple patterns that are rooted at the target BGP. We further consider that a triple pattern is essentially a graph. Thus we propose to leverage the BGP distance proposed in Sect. 2.2.1 of Chap. 3 to form BGP features. For completeness, we briefly introduce the feature modelling here and for the details, please refer to Chap. 3. The eight triple patterns in BGP are mapped to eight distinct graphs and we calculate the BGP distance between the BGPs of two queries using graph edit distance. To build feature vector for a SPARQL query q, we propose to use the BGP distance between q and 18 selected query templates and form a 18-dimensional feature vector using these 18 distances. This method is called template-based feature modelling in Chap. 3 and it outperforms the cluster-based feature modelling by largely reducing the computation time without losing accuracy.

3.3.3 Hybrid Features

We build a hybrid feature by selecting the most predictive features based on algebra and BGP features. Most feature selection approaches rank the candidate features and use this ranking to guide a heuristic search to identify the most predictive features. In this chapter, we will use a similar forward feature selection algorithm, but we choose the contribution to overall prediction performance as the heuristic. The algorithm performs a best-first search in the feature space. It starts with building predictive models using a small number of features and iteratively builds more complex and accurate models by using more features. In each iteration, a new feature is constructed, tested, and added to the feature set. If it improves the overall prediction performance, the feature is selected. We do not consider building multiple models of different types of features for solving the model selection problem. Instead, we use a single type of prediction model, k-NN, because of its excellent performance. Finally, we simply consider the completion of traversing all features as the stopping condition.

Algorithm 1 Plan feature selection algorithm

Require: Training Queries:*data*
Require: Prediction Model:*model*
Require: Feature Models:*feature_models*
Ensure: Prediction performance metric value:*val*
Ensure: Selected feature list:*list*
  1: Initialize: *val* ← *zero*; list ← ∅
  2: **for** *fm* in *feature_models*
  3:    **while** *feature* ←get_feature_next(*fm*,*data*)
  4:    **do**
  5:      *list*=*list*.add(*feature*)
  6:      [*predictions, new_val*]=apply_model(*data*,*list*)
  7:      **if** *new_val*>*val* **then**
  8:        *val* ←*new_val*
  9:      **else**
10:      *list*.remove(*feature*)
11:    **end**
12: **end for**

Algorithm 1 presents the feature selection algorithm. Firstly, we choose from feature described in Sects. 3.3.1 and 3.3.2 (line 2). We put BGP features ahead of algebra features. Then we forwardly choosing single feature from these features and evaluate the performance of prediction with current chosen feature (line 3–6). When new chosen feature contributes to the overall performance, we add it to the candidate list (line 7–8). Otherwise, it is not selected (line 10). The output of the algorithm is a list of selected features.

3.4 Predicting Query Performance

To predict query performance metrics prior to query execution, we apply machine learning techniques on historical executed queries (the training set). We work with query execution time, CPU usage and memory usage as the query performance metrics. Once a prediction model is derived from the training queries, it can then be used to estimate the performance of new requested queries based on the query feature values that can be obtained without executing the query. We train separate prediction models for each of the performance metrics. Our approach does not require prior knowledge of the underlying RDF data, thus it is treated as a black box that the behaviour of queries are learned only from the executed queries.

3.4.1 Predictive Models

We choose two regression approaches in this work, SVR and *k*-NN Regression because SVR is insensitive to outliers [127] and *k*-NN is suitable for irregular data points [90]. The models are trained with features of training queries as well as the actual query performances of these queries. Then the models can be used to estimate the performance of a new issued query through extracting its features.

3.4.1.1 SVR

Four commonly used kernels are considered in our prediction: namely Linear, Polynomial, Radial Basis and Sigmoid, with different kernel parameters γ and r:

- Linear: $K(\mathbf{x}_i, \mathbf{x}_j) = \mathbf{x}_i^{\mathrm{T}} \mathbf{x}_j$.
- Polynomial: $K(\mathbf{x}_i, \mathbf{x}_j) = (\mathbf{x}_i^{\mathrm{T}} \mathbf{x}_j)^r$.
- Radial Basis: $K(\mathbf{x}_i, \mathbf{x}_j) = exp(-\gamma ||\mathbf{x}_i - \mathbf{x}_j||^2), \gamma > 0$.
- Sigmoid: $K(\mathbf{x}_i, \mathbf{x}_j) = tanh(\gamma \mathbf{x}_i^{\mathrm{T}} \mathbf{x}_j) + r$.

3.4.1.2 KNN Regression

We apply four kinds of k-NN regression by considering different weighting schemes.

- *Nearest*. The execution time of the nearest ($k = 1$) neighbour is considered as the predicted time for the new query.
- *Average*. We assign equal weights to each of the k nearest neighbours and use the average of their execution time as the predicted time. The calculation is based on Eq. (3.2) (Sect. 3.2.1).
- *Power*. The weights in *Power* is the power value of weighting scale α. The predicted query time is calculated as follows:

$$t_Q = \frac{1}{k} \sum_{i=1}^{k} (\alpha^i * t_i), \tag{3.3}$$

where α^i is the weight of the i-th nearest query.
- *Exponential*. We apply an exponential decay function with decay scale α to assign weights to neighbours with different distance.

$$t_Q = \frac{1}{k} \sum_{i=1}^{k} (e^{-d_i * \alpha} * t_i), \tag{3.4}$$

where d_i is the distance between target query and its i-th nearest neighbour.

3.4.2 Two-Step Prediction

We observe that prediction process with one-step, i.e., all the training data are input into the model training stage, gives undesirable performance. A possible reason is the fact that our training dataset has queries with various time ranges. Fitting a curve in such irregular data points is often inaccurate. Then we follow a two-step prediction process. We firstly split the training data according to execution time ranges, then we train different prediction for different time ranges. Specifically we put the training queries in four bins (or classes), namely *short*, *medium short*, *medium*, and *long*. The time ranges in these four bins are <0.1 s, 0.1–10 s, 10–3600 s, and >3600 s respectively. We correspondingly label training queries into four labels. Then 12 prediction models are trained, with three models for execution time, CPU usage and memory usage respectively for each of the four classes. When a new query comes, we first classify it to the possible class, then apply the corresponding prediction models of the target class. In this way, the performance improves significantly (see Sect. 3.5.5).

3.5   Experimental Evaluation and Discussion

3.5.1   Setup

Real World Queries   We use real world queries gathered from USEWOD 2014 challenge,[11] which provides query logs from DBPedia's SPARQL endpoint[12] (DBpedia3.9). These logs are formatted in Apache Common Log Format and are encoded. In the data preprocessing step, we process the log files and extract valid queries by decoding, extracting interesting values (IP, date, query string), identifying SPARQL queries from query strings and removing duplicated and invalid queries. Here, invalid queries include all incomplete queries, queries in languages rather than English and queries with syntax errors according to the SPARQL1.1 specification. We work on SELECT queries in the experiments as more than 98% of queries are SELECT queries [84]. We finally retrieve 198,235 valid queries from DBpedia3.9. We randomly choose 10,000 valid queries in our prediction evaluation. We execute these queries 11 times as suggested in [128] and record their execution time, CPU usage and memory usage. We consider the first time as the cold stage and the remaining 10 times as the warm stage. We calculate the average of the remaining 10 times as the actual performance of each query for warm stage prediction. We split the collection to training and test sets according to the 4:1 tradition.

System   The backing system of our local triple store is Virtuoso 7.2, installed on 64-bit Ubuntu 14.04 Linux operation system with 32 GB RAM and 16 CPU cores. We set up a local mirror of DBpedia3.9 English dataset on the Virtuoso server. Table 3.1 shows summary statistics of the dataset. The query performances (execution time,

---

[11]http://usewod.org/.

[12]http://dbpedia.org/sparql/.

Table 3.1 Statistics for
DBpedia3.9 (English)

#.Triples	#.Subject	#.Predicate	#.Object
463,342,557	27,706,241	53,338	133,397,629

CPU usage and memory usage) and query plans are obtained from executing the queries when enabling the profile function of Virtuoso. All the machine learning algorithms are performed on a PC with 64-bit Windows 7, 8 GB RAM and 2.40 GHZ Intel i7-3630QM CPU.

Implementation We use SVR for kernel and linear regression available from LIBSVM [129]. We also use SVM supported by LIBSVM for the classification stage in two-step prediction. k-NN and weighted k-NN regression is designed and implemented using Matlab programming. The heuristic based feature selection algorithm is also implemented in Matlab. The algebra tree used for extracting algebra features is parsed using Apache Jena-2.11.2 library, Java API. Graph edit distance used for building BGP features is calculated using the Graph Matching Toolkit.[13]

Evaluation Metric We follow the suggestion in [118] and use the mean relative error as our prediction metric:

$$relative\ error = \frac{1}{N} \sum_{i=1}^{N} \frac{|actual_i - estimate_i|}{actual_{mean}} \tag{3.5}$$

The difference with the calculation in [118] is that we divide $actual_{mean}$ instead of $actual_i$ because we observe there are zero values for $actual_i$. Relative error is useful when we aim to minimize the relative prediction error for all queries regardless of the actual value. Non-relative error metrics such as square error would be better for minimizing the absolute difference (or its square) in actual and predicted values. One other most widely used metric R^2 is usually computed on the training data [119], but we want to evaluate the fitting of test data.

3.5.2 Prediction Models Comparison

We compared the Linear SVR and SVR with three kernels, namely Polynomial, Radial Basis and Sigmoid. We also calculated the relative error for k-NN when $k = 1$. The feature model used in the experiments was the hybrid feature.

Table 3.2 gives the relative error of predictions of the targeted performance metrics. From the result we can see that the SVR models with various kernels have higher relative errors than k-NN. All the relative errors exceed 97%, indicating the

---

[13]http://www.fhnw.ch/wirtschaft/iwi/gmt.

Table 3.2 Relative Error (%) of prediction on multiple performance metrics

	SVR-L	SVR-P	SVR-R	SVR-S	k-NN ($k = 1$)
Execution time (Cold)	99.69	99.46	99.74	99.68	21.94
Execution time (Warm)	97.59	97.33	97.86	97.57	20.89
CPU usage (Cold)	111.39	110.53	112.25	111.36	38.22
CPU usage (Warm)	106.72	105.46	107.33	106.68	36.25
Memory usage (Cold)	103.39	103.34	103.85	103.41	26.85
Memory usage (Warm)	101.39	101.25	101.93	101.37	23.49

SVR-L denotes SVR-Linear, SVR-P is SVR-Polynomial, SVR-R is SVR-RadialBasis and SVR-S is SVR-Sigmoid

predictions are far from the true values. For cold stage prediction of execution time, k-NN model performs much better with 21.94% in relative error. In warm stage, the relative error goes down to 20.89%. For CPU and memory usage, k-NN model performs much better than SVR models with relative error under 40%, whereas the values exceed 100% using SVR models. We further investigate this result and find two possible reasons. One is that the execution time has a broad range and SVR considers all the data points in the training set to fit the real value, whereas k-NN only considers the points close to the test point. The other reason is we use mean of actual value in Eq. (3.5), and the values that are far from average will lead to distortion of mean value. Given this result, we chose to use k-NN model by default in the following evaluations. Only in the two-step prediction evaluation, we compared SVR with k-NN.

3.5.3   Feature Modelling Comparison

We evaluate the prediction ability of three proposed feature modelling, namely algebra features, BGP features and hybrid features. We further adopt dimensional reduction on hybrid features to evaluate the performance of three most used dimensional reduction algorithms.

3.5.3.1   Performance of Three Types of Features

As the hybrid feature model is built on the feature selection algorithm (see Algorithm 1), we compared its performance with the algebra and BGP feature models to demonstrate the performance comparison with and without feature selection. Figure 3.3 gives the result, showing the relative errors for these three approaches in both cold and warm stages.

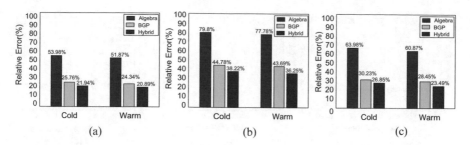

Fig. 3.3 One-step prediction for multiple performances with different features. (**a**) Execution time. (**b**) CPU usage. (**c**) Memory usage

From the figure we can see for all three performance metrics, the hybrid feature performs the best and the BGP feature performs better than the algebra feature. The prediction of execution time gives the best result, with 21.94% relative error in the warm stage and 20.89% relative error in the cold stage. CPU usage is the percentage of CPU used for executing a query. The prediction of CPU usage is slightly poorer. The best prediction is 36.25% relative error when using the hybrid features in the warm stage, and 38.22% is achieved in the cold stage. The reason of prediction on CPU usage having higher relative error is that the CPU scheduling of the underlying operating system for each thread is not the same. Therefore, even for the same query, each time it is executed, the CPU usage might be different.

3.5.3.2 Performance of Dimensional Reduction Algorithms on Hybrid Features

Based on the selected hybrid features, we further applied three feature extraction algorithms to extract the most predictive information. We examine three most used dimension reduction techniques in this work, namely Principle Component Analysis (PCA) [101], Canonical Component Analysis (CCA) [100] and Nonnegative Matrix Factorization (NMF) [130]. We implement them to reduce the dimension of query feature matrix.

Figure 3.4 presents the prediction results for three performance metrics on both warm and cold stages. We observe that NMF shows the worst result, CCA gives medium performance and PCA has the best performance among the three. However, the performance difference between PCA and with or without dimensional reduction is not obvious, indicating that dimension reduction is not suitable for our data. The reason is that dimension reduction algorithms perform well when the original dimension is high, but the dimension of our feature matrices and vectors is relatively low (less than 100). Therefore, we do not apply dimension reduction algorithms in our following evaluations.

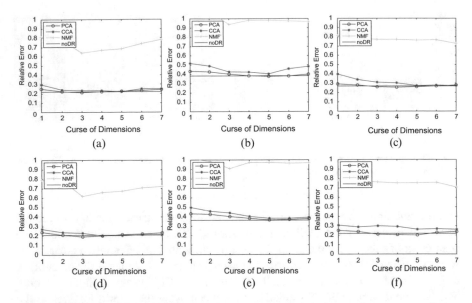

Fig. 3.4 Comparison of dimension reduction algorithms on hybrid features in one-step prediction.
(**a**) Execution time (Cold). (**b**) CPU usage (Cold). (**c**) Memory usage (Cold). (**d**) Execution time
(Warm). (**e**) CPU usage (Warm). (**f**) Memory usage (Warm)

3.5.4   Comparison of Different Weighting Schemes in k-NN Regression

We evaluated three weighting schemes discussed in Sect. 3.4.1.2, namely *Average*,
Power and *Exponential*. All the scaling parameters were chosen through fivefold
cross-validation. We used hybrid features in this evaluation. Both warm and cold
stages were evaluated. We presented only the result for execution time because
other performance metrics provide similar results. We observe from Fig. 3.5 that
the power weighting achieves the best performance. In the warm stage, the 15.32%
relative error is achieved when $k = 5$. The trend of relative error returns to upward
after $k = 5$. *Average* is the worst weighting method for our data. Exponential
weighting does not perform as well as we expected although it is better than
average weighting. Weighting schemes show similar performances when the query
execution is in the cold stage, i.e., when $k = 5$, the power weighting achieves
the lowest relative error of 18.29%. We therefore used $k = 5$ power weighting
in following evaluations.

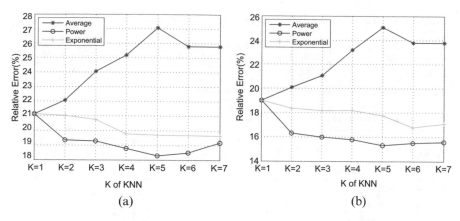

Fig. 3.5 One-step execution time prediction of different weighting method on k-NN. (**a**) Cold. (**b**) Warm

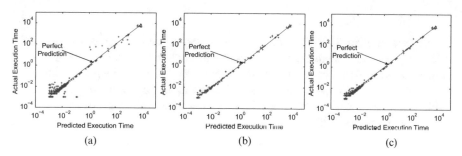

Fig. 3.6 One-step and two-step prediction fitting for execution time (s). (**a**) One-step (Warm). (**b**) Two-step (Cold). (**c**) Two-step (Warm)

3.5.5 Performance of Two-Step Prediction

We observe that the performance of one-step prediction is not desirable with the lowest relative error 15.32% when predicting execution time for warm stage querying. A possible reason is the broad execution time range of actual SPARQL queries. The long time queries will distort the mean of actual observations that make the relative error inaccurate. We thus propose the two-step prediction process as discussed in Sect. 3.4.2. Here we evaluated the performance of two-step prediction. We used Support Vector Machine for the classification task in two-step prediction and achieved accuracy of 98.36%, indicating that we can accurately predict the time range in the first step. For comparison, we used log-log plotting for the one-step and two-step prediction in Fig. 3.6. We only show the results for execution time in both cold and warm stages here. The one-step warm stage prediction is worse than the two-step cold stage prediction and the performance gap to two-step warm stage prediction is larger.

Table 3.3 Relative errors (%) of two-Step prediction with k-NN and SVR

Prediction model	Execution time (s)	CPU usage	Memory usage
5NN($\alpha = 0.3$)	11.06, 9.81	37.78, 35.34	23.58, 18.94
SVR-Polynomial	22.39, 20.30	60.15, 58.56	40.34, 36.62

The values delimited by comma are for (cold, warm) respectively

Table 3.4 Comparison to state-of-the-art work

Models	Our approach SVN + Weighted KNN	Approach in [90] X-means + SVM + SVR
Training time 1k queries (s)	51.36	1548.45
err% of execution time	9.81	14.39
err% of CPU usage	35.34	38.67
err% of memory usage	18.94	22.24

Training time for 1000 queries are compared as well as the relative errors for each performance metric

We also applied SVR-Polynomial in two-step prediction for comparison. We selected polynomial here because it shows the best performance among all the kernels we considered in this work (see Table 3.2). We used the power weighting of k-NN with $k = 5$ and $\alpha = 0.3$ as it shows the best performance (Sect. 3.5.4). Table 3.3 shows the result of comparison. It shows that SVR regression model still does not perform desirably.

3.5.6 Comparison to State-of-the-Art

The only work that exploits machine learning algorithms to predict SPARQL query is the work in [90]. In our evaluation, we implemented their approach and compared their work with ours. To implement the approach in [90], we calculated GED for all pairs of 1000 randomly chosen queries and cluster the training queries using X-means per description in [90], then built distance feature vectors for each test query, and each instance is the distance between the test query and the center query of each cluster. We then trained prediction models for each cluster. Finally we used the trained model to predict the performance of test queries. Table 3.4 shows the result of comparison on warm stage querying. The training time includes feature modelling, clustering and classification for work in [90]. The first part takes most time because the calculation of GED for all training queries is time-consuming. In our approach, we reduce the GED calculation dramatically. But this calculation still takes most of the time in the prediction process. The time gap of training process between ours and theirs will be enlarged when more training queries are involved because their approach takes squared time. We do not have a clustering process that further reduces the time. Our approach also shows better prediction performance with lower relative error for all three prediction performances.

3.6 Discussions

In this section, we discuss the observations and issues from our experience in this work.

Plan Features Some works use machine learning techniques that leverage the information provided by the query plan, which is given by a query optimizer. The information that the plan provides in these works includes estimated total execution time, estimated result row count, estimated time and row count for each operators. However, there are two obstacles for using such information in our work. Firstly, this information is based on the cost model estimation, which has been proven ineffective [118, 119]. It is unlikely to achieve desirable performances based on inaccurate estimations. Secondly, as we target open source triple stores to benefit more data consumers, we are only able to obtain information from these systems. However, most of them fail to provide an explicit query plan. Thus we turn to choose structure-based features that can be obtained directly from query text. As argued in many works, similarly structured queries may have huge performance gaps due to the irregular distribution of values in the queried data. But from our practical experience in this work, we observed that although it leads to distortion of the prediction, the error rate is acceptable based on the limited features we can acquire.

Cost Model vs Machine Learning Cost model based optimization estimates execution time based on the cost. Although it is arguably less accurate than machine learning based prediction, it is faster than machine learning prediction. The downside is that those estimations are always inaccessible for personal users. To promote the usability of Semantic Web techniques and RDF data, it is better to be more consumer friendly. Therefore, the machine learning approaches are a better choice, as the publicly accessible tools are easy to use for training and testing.

Training Size In the training process, the larger the size of the training data, the better performance we can get. The reason is that more data variety is seen and the model will be less sensitive to unforeseen queries. However, in practice, it is time consuming to obtain the query execution time of a large collection of queries. That is the possible reason why many other works only use small sizes of queries in their evaluation. This fact will cause the bias of the prediction result and makes similar works hard to compare. Although our experimental query set is larger than theirs, we still consider to further enlarge the size of our query set to cover more various queries in the future.

Dynamic vs Static Data In dynamic query workloads, the queried data is frequently updated. Therefore, the prediction might perform poorly due to lack of update of training data. Our work focuses on prediction on static data and we expect training to be done in a periodical manner. In the future we plan to investigate the techniques that can make prediction more available for continuous retraining, which reflects recently executed queries.

Moreover, the query performance would vary when execution environment changes. Thus we perform the evaluations on fixed dataset and run 10 times to get the warm stage performance. Our approach shows desirable prediction performance in this scenario. Although our approach is not designed to predict query execution performance under changing environments (e.g., updating of data, changing of resources etc.), it can be an indicator of the query performances compared to other queries.

3.7 Related Work

Although there are very limited previous works that pertain to predicting query performance via machine learning algorithms in the context of SPARQL queries, the literature of the approaches and ideas presented in this chapter is extensive. In this section, we introduce some representative works that are closely related to our work.

3.7.1 Query Performance Prediction via Machine Learning Algorithms

Predicting query execution time by leveraging machine learning techniques has recently gained significant interest in the database research community. Akdere et al. [118] propose to predict the execution time using Support Vector Machine (SVM). They build predictors by leveraging query plans provided by the PostgreSQL optimizer. The authors also choose operator-level predictors and then combine the two with heuristic techniques. The work studies the effectiveness of machine learning techniques for predicting query latency of both static and dynamic workload scenarios. Ganapathi et al. [119] consider the problem of predicting multiple performance metrics at the same time. The authors also choose query plans to build the feature matrix. Kernel Canonical Correlation Analysis (KCCA) is leveraged to build the prediction model as it is able to correlate two high-dimensional datasets. As addressed by the authors, it is hard to find a reverse mapping from feature space back to the input space and they consider the performance metric of k-nearest-neighbours to estimate the performance of target query. Li et al. [116] estimate CPU time and I/O of a query execution plan. The work also addresses the problem of robust estimation for queries that are not observed in the training stage.

3.7.2 SPARQL Query Optimization

RDBMS-based triple stores rely on the query optimization techniques of the relational database systems to evaluate SPARQL queries. Recent works focus on optimizing the joins of SPARQL queries [128, 131]. Neumann et al. [131] introduce characteristic sets, which works with dynamic programming algorithm to provide more accurate selectivity estimations for star-like SPARQL queries. Gubichev et al. [128] propose a SPARQL-tailored join-ordering algorithm aiming at large SPRAQL queries. RDF native query engines use rule-based optimization or leverage heuristics and statistics about the data for selecting efficient query plans [122]. These approaches generally work without any knowledge of the underlying data. Quilitz et al. [132] apply a set of logical rules to a query engine, to calculate all equivalent query plans for a given query and then choose the most optimized query plan to be executed. Stocker et al. [122] present optimization techniques with pre-computed statistics. Based on the statistics, triple patterns are reordered before execution to achieve efficient query processing. Tsialiamanis et al. [121] propose a heuristics-based SPARQL optimizer to maximize the number of merge-joins to speed up the query processing.

3.8 Summary

In this chapter, we have leveraged machine learning techniques to predict multiple performance metrics for SPARQL queries. We transformed a given SPARQL query to a vector representation. Feature vectors are built by exploiting the syntactic and structural characteristics of SPARQL queries. SVR and KNN regression models are adopted as prediction models in this work. We observed that KNN performs better than SVR for our data because of the irregular distribution of query performance. The dimension reduction technique is not suitable for our low-dimension feature matrices and vectors. The proposed two-step prediction performs much better because it considers the broad range of observed execution time. The prediction of execution time is more accurate, however for CPU usage, the prediction is not desirable. The reason is that the CPU usage is rarely consistent even for identical queries executed in the warm stage. The prediction in the warm stage is generally better than in the cold stage. We observed that the reason comes from same structured queries. We also observed that many queries are issued by programmatic users, who tend to issue queries using query templates.

We have discussed two topics of querying curated KBs, namely predicting and improving querying performances in Chap. 2 and this chapter. From the next chapter, we will discuss our methods on solving several knowledge extraction issues in open KBs.

Chapter 4
An Efficient Knowledge Clustering Algorithm

In this chapter, we discuss our solution for orthogonal Non-negative Matrix Factorization (ONMF), which is a promising solution for the desirable performance in clustering applications. In ONMF, one factor matrix is posed orthogonal constraint. Our proposed method is called Non-linear *Riemannian* Conjugate Gradient ONMF (NRCG-ONMF), in which the orthogonal factor matrix is given orthogonality constraint through the searching on *Stiefel* manifold. Extensive experiments on both synthetic and real-world data sets show consistent superiority of our method over other approaches in terms of orthogonality preservation, convergence speed and clustering performance. This chapter is based on our research reported in [133].

4.1 Overview of Clustering with Non-negative Matrix Factorization

Given a matrix, Non-negative Matrix Factorization (NMF) aims to find two non-negative factor matrices whose product approximates that matrix. This enhances the interpretability initiated by Paatero and Tapper [134] as Positive Matrix Factorization and popularized by Lee and Seung [130]. NMF has enjoyed much success in text mining [135], image processing [136], recommendation systems [137] and many other areas, and has attracted much theoretical and practical attention.

Orthogonal NMF (ONMF), first introduced by Ding et al. [138], is a variant of NMF with an additional orthogonal constraint on one of the factor matrices. Without loss of generality, the problem can be written as follows:

$$\min_{\mathbf{U},\mathbf{V}} ||\mathbf{X} - \mathbf{U}\mathbf{V}^{\mathsf{T}}||_F^2, \text{ s.t. } \mathbf{U}^{\mathsf{T}}\mathbf{U} = \mathbf{I}_r, \mathbf{U} \geq \mathbf{0}, \mathbf{V} \geq \mathbf{0}, \tag{4.1}$$

where $\mathbf{X}$ is an $m \times n$ matrix, $\mathbf{U} \in \mathbb{R}^{m \times r}$ and $\mathbf{V} \in \mathbb{R}^{n \times r}$ are two factor matrices, and $|| \cdot ||_F$ is the *Frobenius* norm. In Problem (4.1), $\mathbf{I}_r$ is an identity matrix

© Springer International Publishing AG, part of Springer Nature 2018
W. E. Zhang, Q. Z. Sheng, *Managing Data From Knowledge Bases:
Querying and Extraction*, https://doi.org/10.1007/978-3-319-94935-2_4

of size $r \times r$. The factor matrix $\mathbf{U}$ is imposed on the orthogonal constraint. ONMF has been shown to be identical to k-means when the *Frobenius* norm is used as divergence/distance [138]. Thus one factor matrix is corresponding to the cluster centers and the other matrix is corresponding to the cluster membership indicators [139]. The orthogonality on the columns of $\mathbf{U}$ is identical to clustering rows of the input data matrix $\mathbf{X}$ [140] and makes the clusters more distinct than without orthogonal constraint [141]. Moreover, the orthogonal constraint reduces computation complexity of NMF approaches [142]. Due to these properties, ONMF becomes increasingly popular in clustering tasks [140]. Most existing ONMF methods either enforce the orthogonality directly on the factor matrix [138, 143] as constraints or in the objective function [141]. In addition, the authors in [135] propose a projection gradient method leveraging the manifold constraint. Others have recently tried to approximate ONMF problem by the Non-negative Principle Component Analysis (NNPCA) problem [144] gaining good results. However these methods do not preserve orthogonality well (except for the work in [135]) and/or suffer from slow convergence (Sect. 4.3.4.2).

Another branch of approaches to improve clustering interpretation on NMF is to relax the non-negativity constraint on one of the factor matrices. Ding et al. [145] refer to this constrained NMF as Semi-NMF and prove its applicability in the clustering perspective where one factor matrix represents the cluster centroids and the other represents soft membership indicators for every data point.

However, limited works consider both orthogonality and relaxed non-negativity to improve the clustering performance of NMF. In this work, we formalize the problem as follows:

$$\min_{\mathbf{U},\mathbf{V}} ||\mathbf{X} - \mathbf{U}\mathbf{V}^\mathsf{T}||_F^2 + \frac{\lambda}{2}||\mathbf{U}\mathbf{V}^\mathsf{T}||_F^2, \text{ s.t. } \mathbf{U}^\mathsf{T}\mathbf{U} = \mathbf{I}_r, \mathbf{V} \geq \mathbf{0}, \qquad (4.2)$$

where $\mathbf{U}$ is with orthogonality constraint only and λ is the regularization parameter. The regularizer $\frac{1}{2}||\mathbf{U}\mathbf{V}^\mathsf{T}||_F^2$ is introduced to avoid over-fitting issue.

To solve Problem (4.2), we propose a Nonlinear *Riemannian* Conjugate Gradient ONMF (NRCG-ONMF), which updates $\mathbf{U}$ and $\mathbf{V}$ alternatively and iteratively: when updating $\mathbf{U}$ (resp., $\mathbf{V}$), $\mathbf{V}$ (resp., $\mathbf{U}$) is fixed. Specifically, when updating $\mathbf{U}$, a Nonlinear *Riemannian* Conjugate Gradient method (NRCG) is proposed. This method preserves the orthogonality of $\mathbf{U}$ in the setting of *Stiefel* manifold which is an embedded sub-manifold of the *Riemannian* manifold [146]. The NRCG method performs nonlinear search on *Stiefel* manifold by three steps: (1) identifying the search direction on the tangent space of *Stiefel* manifold regarding $\mathbf{U}$ with Conjugate Gradient (CG) rules, (2) moving the point along the search direction with carefully determined step size, and (3) projecting the new point back to the *Stiefel* manifold to preserve the orthogonality with a *Retraction* operation. When updating $\mathbf{V}$, due to the nonnegative constraint, we update each entry of $\mathbf{V}$ separately in a coordinate descent manner by closed-form solutions.

The main contributions of our work are summarized as follows.

- We exploit the orthogonality and semi-nonnegativity constraints in NMF aiming at improving the clustering performance. We develop an efficient nonlinear *Riemannian* conjugate gradient method with Barzilai-Borwein (BB) step size to update **U** in order to preserve the orthogonality of **U** with fast convergence speed.
- Based on the NRCG method, we propose a NRCG-ONMF method to address Problem (4.2), where we update **U** and **V** alternatively. The convergence of NRCG-ONMF is also analyzed.
- Extensive experiments on both synthetic and real-world data sets show that our method outperforms the related works in terms of orthogonality, convergence speed and clustering performance.

The rest of this chapter is organized as follows. We introduce our proposed algorithm in Sect. 4.2. The experiments are presented in Sect. 4.3 followed by related works in Sect. 4.4 and summary in Sect. 4.5.

4.2 Orthogonal Non-negative Matrix Factorization Over Stiefel Manifold

4.2.1 Notations

Throughout the paper, we use bold uppercase and lowercase letters to represent the matrices and vectors respectively. We denote by the superscript $^\top$ the transpose of a vector/matrix. The *Frobenius* norm of **X** is defined as $||\mathbf{X}||_F$. The list of main notations can be found in Table 4.1.

4.2.2 Optimization on Stiefel Manifold

To begin with, we introduce some geometries regarding *Stiefel* manifold. The *Stiefel* manifold, denoted by St_r^m, is defined as

Table 4.1 Notation conventions

Notations	Descriptions
U, V, G, X, Z, I, E	Matrices
$\boldsymbol{\eta}_k, \mathbf{x}_i, \mathbf{b}_j, \mathbf{v}_i$	Vectors
$\mathcal{M}_r, \mathrm{St}_r^m$	Manifolds
$T_{\mathbf{U}}\mathcal{M}_r$	Tangent spaces
$\lambda, \alpha_k, \beta_k, \delta, \epsilon, w_i, s_{jl}$	Parameters
m, n, r	Integers
$F(\mathbf{U}, \mathbf{V}), f(\mathbf{U}), g(\mathbf{V})$	Functions

$$\mathrm{St}_r^m = \{\mathbf{U} \in \mathbb{R}^{m \times r} | \mathbf{U}^\mathsf{T}\mathbf{U} = \mathbf{I}_r\},$$

which is a set of m by r orthonormal matrices [146]. The *Stiefel* manifold is compact, and its dimension is $(mr - \frac{1}{2}r(r+1))$. For convenience of presentation, hereafter, we denote St_r^m by $\mathcal{M}_r$ which is a nonlinear manifold. The tangent space to $\mathcal{M}_r$ at $\mathbf{U}$, denoted by $T_\mathbf{U}\mathcal{M}_r$, is the set of all tangent vectors to $\mathcal{M}_r$ at $\mathbf{U}$. To be more specific, the tangent space of *Stiefel* manifold $\mathcal{M}_r$ at $\mathbf{U}$ is given as [146]:

$$
\begin{aligned}
T_\mathbf{U}\mathcal{M}_r &:= \left\{ \mathbf{Z} \in \mathbb{R}^{m \times r} : \mathbf{Z}^\mathsf{T}\mathbf{U} + \mathbf{U}^\mathsf{T}\mathbf{Z} = \mathbf{0} \right\} \\
&= \left\{ \mathbf{U}\mathbf{K} + (\mathbf{I}_m - \mathbf{U}\mathbf{U}^\mathsf{T})\mathbf{J} : \mathbf{K} = -\mathbf{K}^\mathsf{T} \in \mathbb{R}^{r \times r}, \mathbf{J} \in \mathbb{R}^{m \times r} \right\}.
\end{aligned}
\tag{4.3}
$$

On the tangent space $T_\mathbf{U}\mathcal{M}_r$ for any $\mathbf{U} \in \mathcal{M}_r$, we introduce the standard inner product as a metric:

$$\langle \mathbf{Y}, \mathbf{Z} \rangle_\mathbf{U} := \mathrm{tr}(\mathbf{Y}^\mathsf{T}\mathbf{Z}), \ \forall \mathbf{Y}, \mathbf{Z} \in T_\mathbf{U}\mathcal{M}_r. \tag{4.4}$$

Then, we can view St_r^m (i.e. $\mathcal{M}_r$) as a sub-manifold of *Riemannian* manifold $\mathbb{R}^{m \times r}$. Given a smooth function $f(\mathbf{U})$ on $\mathcal{M}_r$, its *Riemannian* gradient is given as the orthogonal projection onto the tangent space of the gradient of $f(\mathbf{U})$. Specifically, the orthogonal projection of any $\mathbf{Z} \in \mathbb{R}^{m \times r}$ onto the tangent space at $\mathbf{U}$ is defined as [146]:

$$
\begin{aligned}
P_{T_\mathbf{U}\mathcal{M}_r}(\mathbf{Z}) :&= \mathbf{U}\left(\frac{\mathbf{U}^\mathsf{T}\mathbf{Z} - \mathbf{Z}^\mathsf{T}\mathbf{U}}{2} \right) + (\mathbf{I}_m - \mathbf{U}\mathbf{U}^\mathsf{T})\mathbf{Z} \\
&= \mathbf{Z} - \mathbf{U}\mathrm{Sym}(\mathbf{U}^\mathsf{T}\mathbf{Z}),
\end{aligned}
\tag{4.5}
$$

where $\mathrm{Sym}(\mathbf{A}) := \frac{1}{2}(\mathbf{A} + \mathbf{A}^\mathsf{T})$. Let $\mathbf{G}_\mathbf{U} = \nabla f(\mathbf{U})$ be the gradient of $f(\mathbf{U})$, the *Riemannian* gradient of $f(\mathbf{U})$ on $\mathcal{M}_r$, denoted by $\mathrm{grad} f(\mathbf{U})$, can be calculated using:

$$
\begin{aligned}
\mathrm{grad} f(\mathbf{U}) &= P_{T_\mathbf{U}\mathcal{M}_r}(\mathbf{G}_\mathbf{U}) \\
&= \mathbf{G}_\mathbf{U} - \mathbf{U}\mathrm{Sym}(\mathbf{U}^\mathsf{T}\mathbf{G}_\mathbf{U}).
\end{aligned}
\tag{4.6}
$$

Regarding our problem in (4.2), the gradient $\mathbf{G}_\mathbf{U}$ can be computed by

$$\mathbf{G}_\mathbf{U} = (\mathbf{U}\mathbf{V}^\mathsf{T} - \mathbf{X})\mathbf{V} + \lambda \mathbf{U}\mathbf{V}^\mathsf{T}\mathbf{V}. \tag{4.7}$$

By applying Eqs. (4.6) and (4.7), $\mathrm{grad} f(\mathbf{U})$ can be computed by

$$\mathrm{grad} f(\mathbf{U}) = \frac{1}{2}\mathbf{U}\mathbf{V}^\mathsf{T}\mathbf{X}^\mathsf{T}\mathbf{U} - \frac{1}{2}\mathbf{X}\mathbf{V}. \tag{4.8}$$

For convenience, we define the objective function as

$$F(\mathbf{U}, \mathbf{V}) = \frac{1}{2}||\mathbf{X} - \mathbf{U}\mathbf{V}^{\mathsf{T}}||_F^2 + \frac{\lambda}{2}||\mathbf{U}\mathbf{V}^{\mathsf{T}}||_F^2. \tag{4.9}$$

So Problem (4.2) can be written as

$$\min_{\mathbf{U},\mathbf{V}} F(\mathbf{U}, \mathbf{V}), \quad \text{s.t.,} \ \mathbf{U}^{\mathsf{T}}\mathbf{U} = \mathbf{I}_k, \mathbf{V} \geq \mathbf{0}.$$

To address this problem, we propose the NRCG-ONMF method as in Algorithm 1, which NRCG-ONMF updates $\mathbf{U}$ and $\mathbf{V}$ alternatively until reaching the stopping criteria. The details of updating $\mathbf{U}$ and $\mathbf{V}$ are discussed in following sections. In Sect. 4.2.3, we introduce the NRCG, a *Riemannian* optimization scheme, for updating the orthogonal factor matrix $\mathbf{U}$. We discuss the closed-form update of the non-negative factor matrix $\mathbf{V}$ in Sect. 4.2.4. The convergence analysis of NRCG-ONMF is given in Sect. 4.2.5.

4.2.3 Update U via NRCG

When updating $\mathbf{U}$, the variable $\mathbf{V}$ is fixed, and we equivalently address the following optimization problem:

$$\min_{\mathbf{U}} f(\mathbf{U}), \quad \text{s.t.,} \ \mathbf{U}^{\mathsf{T}}\mathbf{U} = \mathbf{I}_k, \tag{4.10}$$

where $f(\mathbf{U}) = \frac{1}{2}||\mathbf{X} - \mathbf{U}\mathbf{V}^{\mathsf{T}}||_F^2 + \frac{\lambda}{2}||\mathbf{U}\mathbf{V}^{\mathsf{T}}||_F^2$ with $\mathbf{V}$ being fixed and $\mathbf{U}$ being restricted on the *Stiefel* manifold. Problem (4.10) is a non-convex and nonlinear optimization problem. To address it, we develop a NRCG method using BB step size for minimizing $f(\mathbf{U})$.

On Euclidean space, the classical CG method does the search with two major steps, i.e., *finding a conjugate search direction* and *determining the step size*. However, unlike classical CG method, when doing the optimization on the *Riemannian* manifold, two additional geometric operations, namely *Vector Transport* and *Retraction*, are required.

Algorithm 1 NRCG-ONMF

Given $\mathbf{X}$, initialize $\mathbf{U}_0, \mathbf{V}_0$. Set $k = 1$.
1: Update $\mathbf{U}$ by NRCG method in Sect. 4.2.3.
2: Update $\mathbf{V}$ according to Algorithm 3 in Sect. 4.2.4.
3: Quit if stopping conditions achieve.
4: Let $k = k + 1$ and go to step 1.

4.2.3.1   Conjugate Gradient Descent on $\mathcal{M}_r$

Similar to gradient based optimization methods on the Euclidean space, at any point $\mathbf{U}$, the optimization on *Stiefel* manifold requires identifying a search direction, which is a tangent vector to $\mathcal{M}_r$ at $\mathbf{U}$. A direct choice would be the steepest descent direction, i.e., the negative of the gradient of the objective function. However, the steepest descent method may incur slow convergence speed. To avoid this, we seek to apply the conjugate search direction.

On the Euclidean space $\mathbb{R}^n$, the conjugate search direction $\boldsymbol{\eta}_k$ in nonlinear CG is calculated by

$$\boldsymbol{\eta}_k = -\mathrm{grad}\, f(\mathbf{U}_k) + \beta_k \boldsymbol{\eta}_{k-1}, \tag{4.11}$$

where the initial direction $\boldsymbol{\eta}_0$ is set to the steepest descent direction, and β_k is calculated by, for example, the Hestenes-Stiefel's rule (HS) [147, 148] as follows:

$$\beta_k = \frac{\langle \mathrm{grad}\, f(\mathbf{U}_k), \mathrm{grad}\, f(\mathbf{U}_k)\rangle - \langle \mathrm{grad}\, f(\mathbf{U}_k), \mathrm{grad}\, f(\mathbf{U}_{k-1})\rangle}{\langle \boldsymbol{\eta}_{k-1}, \mathrm{grad}\, f(\mathbf{U}_k)\rangle - \langle \boldsymbol{\eta}_{k-1}, \mathrm{grad}\, f(\mathbf{U}_{k-1})\rangle}. \tag{4.12}$$

Different from methods on the Euclidean space, the search direction on a manifold is adapted to follow a path on the manifold [146]. Since $\mathrm{grad}\, f(\mathbf{U}_k) \in T_{\mathbf{U}_k}\mathcal{M}_r$, $\mathrm{grad}\, f(\mathbf{U}_{k-1}) \in T_{\mathbf{U}_{k-1}}\mathcal{M}_r$, and $\boldsymbol{\eta}_{k-1}$ are in different tangent spaces of the manifold, Eq. (4.11) is not applicable on *Riemannian* manifolds. To address this, we need to introduce the *Vector Transport*.

4.2.3.2   Vector Transport

The *Vector Transport* $\mathscr{T}$ on a manifold $\mathcal{M}_r$ is a smooth mapping that transports tangent vectors from one tangent space to another [149]. Specifically, let $\mathscr{T}_{\mathbf{U}\rightarrow\mathbf{Y}}(\boldsymbol{\zeta}_{\mathbf{U}})$ denote the transport from one tangent space $T_{\mathbf{U}}\mathcal{M}_r$ to another tangent space $T_{\mathbf{Y}}\mathcal{M}_r$, where $\boldsymbol{\zeta}_{\mathbf{U}}$ denotes the tangent vector on $T_{\mathbf{U}}\mathcal{M}_r$, the conjugate direction can be calculated by

$$\boldsymbol{\eta}_k = -\mathrm{grad}\, f(\mathbf{U}_k) + \beta_k \mathscr{T}_{\mathbf{U}_{k-1}\rightarrow\mathbf{U}_k}(\boldsymbol{\eta}_{k-1}), \tag{4.13}$$

where β_k can be computed according to the HS rule (Eq. (4.12)). Here, $\mathrm{grad}\, f(\mathbf{U}_k)$ is the *Riemannian* gradient of $f(\mathbf{U}_k)$ on the *Riemannian* manifold $\mathcal{M}_r$, which can be computed by Eq. (4.8). For the calculation of $\mathscr{T}_{\mathbf{U}\rightarrow\mathbf{Y}}(\boldsymbol{\zeta}_{\mathbf{U}})$, we adopt the method in [149] and readers are referred to this paper for details.

Given a search direction $\boldsymbol{\eta}_k$ at the k-th iteration, one may move the point $\mathbf{U}_k$ along the search direction to a new point $(\mathbf{U}_k + \alpha \boldsymbol{\eta}_k)$, where α is the step size. However, this new point will no longer stay on the manifold when $\alpha > 0$. To address this, we need to introduce the *Retraction* operation.

4.2.3.3 Retraction

Retraction is a projection mapping from the tangent bundle onto the manifold to keep the new point on the manifold [146]. In other words, with the retraction mapping, one can move points in the direction of a tangent vector and stay on the manifold. Given any $\boldsymbol{\xi} \in T_{\mathbf{U}}\mathcal{M}_r$, the retraction on *Stiefel* manifold $\mathcal{M}_r$, denoted by $R_{\mathbf{U}}(\boldsymbol{\xi})$, can be computed as

$$R_{\mathbf{U}}(\boldsymbol{\xi}) := qf(\mathbf{U} + \boldsymbol{\xi}), \quad s.t. \ \boldsymbol{\xi} \in T_{\mathbf{U}}\mathcal{M}_r, \tag{4.14}$$

where $qf(\mathbf{A})$ denotes the Q factor of the decomposition of $\mathbf{A}$ as $\mathbf{A} = \mathbf{QR}$ [146]. Apparently, we have $R_{\mathbf{U}}(\boldsymbol{\xi}) \in \mathcal{M}_r$, which is set as the new point. Given a search direction $\boldsymbol{\eta}_k$ at $\mathbf{U}_k$, set $\boldsymbol{\xi} = \alpha\boldsymbol{\eta}_k$, then the retraction can be performed by

$$\mathbf{U}_{k+1} = R_{\mathbf{U}_k}(\alpha\boldsymbol{\eta}_k) := qf(\mathbf{U}_k + \alpha\boldsymbol{\eta}_k), \tag{4.15}$$

where α is the step size to be determined.

4.2.3.4 Determination of the Step Size

A good step size would guarantee the convergence of a search algorithm and accelerate its speed without increasing much cost. Many search rules (e.g., Armijo-Wolfe rule [150]) have been proposed to find a suitable step size along a given search direction. In this work, we choose the BB step size [151] as it can significantly reduce the total number of iterations through empirical studies on optimization problems that subject to spherical constraints [152].

BB adjusts the step size α by considering second order information (similar to Newton method) but without computing the second derivative of objective function. We adopt a non-monotone line search method from [153], where α_k satisfies non-monotone Wolfe conditions:

$$f(R_{\mathbf{U}_k}(\alpha_k\boldsymbol{\eta}_k)) \leq C_k + \delta\alpha_k \langle \text{grad} f(\mathbf{U}_k), \boldsymbol{\eta}_k \rangle, \tag{4.16}$$

and

$$C_{k+1} = (\sigma Q_k C_k + f(\mathbf{U}_{k+1}))/Q_{k+1}, \tag{4.17}$$
$$Q_{k+1} = \sigma Q_k + 1,$$

where $C_0 = f(\mathbf{U}_0)$ and $Q_0 = 1$. Once the condition in (4.16) is fulfilled, set $\alpha_{k+1} = \tau\alpha_k$. The parameters $\delta, \sigma, \tau \in (0, 1)$. Recall $\boldsymbol{\eta}_k \in T_{\mathbf{U}_k}\mathcal{M}_r$ is the search direction. The existence of α_k is guaranteed according to the following lemma.

Lemma 4.1 *Let* $\mathbf{U}_k \in \mathcal{M}_r$, *and* $\boldsymbol{\zeta}_k \in T_{\mathbf{U}_k}\mathcal{M}_r$ *be a descent direction. Then there exists an* α_k *that satisfies the condition in (4.16).*

Proof Recall that $\mathcal{M}_r$ is a compact manifold. Since $\boldsymbol{\zeta}_k$ is a descent direction, it follows that $\mathbf{0} \neq \operatorname{grad} f(\mathbf{U}_k)$, which implies that $\langle \operatorname{grad} f(\mathbf{U}_k), \boldsymbol{\zeta}_k \rangle < 0$. Since $C_k \geq f(\mathbf{U}_k)$ (see [153]) and $R_{\mathbf{U}_k}(\alpha\boldsymbol{\eta}_k))$ is continuous in α, there must exist an $\widehat{\alpha}$ such that the inequality in (4.16) holds $\forall \alpha \in (0, \widehat{\alpha}]$.

The update of $\mathbf{U}$ is summarized in Algorithm 2.

4.2.4   Update V

When updating $\mathbf{V}$, we fix $\mathbf{U}$ as a constant. Due to the non-negative constraint of $\mathbf{V}$, updating all the elements in $\mathbf{V}$ is difficult. Therefore, we adopt a coordinate-decent update for $\mathbf{V}$, namely we update each entry of $\mathbf{V}$ in a random order. For example, to update V_{jl}, we have

$$(\mathbf{U}, \mathbf{V}) \rightarrow (\mathbf{U}, \mathbf{V} + s_{jl}E_{jl}), \tag{4.18}$$

where $\mathbf{E}$ is a $n \times r$ matrix with all elements zero except that $E_{j,l} = 1$, and s_{jl} is regarded as the step size when updating V_{jl}. The problem of finding s_{jl} in (4.18) can be cast as the following optimization problem:

$$\min_{s_{jl}:\mathbf{V}_{jl}+s_{jl}\geq 0} g_{jl}(s_{jl}) = g(\mathbf{V} + s_{jl}E_{jl}), \tag{4.19}$$

where $g(\mathbf{V}) = \frac{1}{2}||\mathbf{X} - \mathbf{U}\mathbf{V}^\mathsf{T}||_F^2 + \frac{\lambda}{2}||\mathbf{U}\mathbf{V}^\mathsf{T}||_F^2$ with $\mathbf{U}$ being fixed. Similar to [154], we can rewrite $g_{jl}(s_{jl})$ as

$$g_{jl}(s_{jl}) = \frac{1}{2}g_{jl}''(0)s_{jl}^2 + g_{jl}'(0)s_{jl} + g_{jl}(0), \tag{4.20}$$

Algorithm 2 Nonlinear Riemannian Conjugate Gradient (NRCG) update for U

1: Given $\mathbf{U}_k, \mathbf{X}, \mathbf{V}_k, \alpha > 0, \delta, \sigma, \tau \in (0, 1), k \geq 1$.
2: Compute the gradient $\operatorname{grad} f(\mathbf{U}_k)$ by (4.8).
3: Compute a conjugate direction $\boldsymbol{\eta}_k$ according to (4.13).
4: Choose a BB step size α_k to satisfy conditions (4.16) and (4.17), and set $\mathbf{U}_{k+1} = R_{\mathbf{U}_k}(\alpha_k\boldsymbol{\eta}_k)$.
5: Output $\mathbf{U}_{k+1}$ in order to update $\mathbf{V}$.

---

where the $g'_{jl}(0)$ and $g''_{jl}(0)$ denote the first derivative and second derivative of $g_{jl}(s_{jl})$ at $s_{jl} = 0$, respectively. It follows that

$$
\begin{aligned}
g'_{jl}(0) &= \left(\frac{\partial g}{\partial \mathbf{V}}\right)_{jl} \\
&= \left((\mathbf{UV}^\mathsf{T} - \mathbf{X})^\mathsf{T}\mathbf{U} + \lambda(\mathbf{UV}^\mathsf{T})^\mathsf{T}\mathbf{U}\right)_{jl} \\
&= (1 + \lambda)V_{jl} - (\mathbf{X}^\mathsf{T}\mathbf{U})_{jl},
\end{aligned}
\tag{4.21}
$$

and

$$
g''_{jl}(0) = \left(\frac{\partial^2 g}{\partial \mathbf{V}^2}\right)_{jl} = (1 + \lambda).
\tag{4.22}
$$

By ignoring the non-negative constraint, one can get a closed form minimizer of $g_{jl}(s_{jl})$ as follows

$$
s_{jl} = -\frac{g'_{jl}(0)}{g''_{jl}(0)}.
\tag{4.23}
$$

Accordingly, considering the non-negative constraint on $\mathbf{V}$, the computation of s_{jl} is modified as

$$
s^*_{jl} = \max\left(0, \, V_{jl} - \frac{g'_{jl}(0)}{g''_{jl}(0)}\right) - V_{jl}.
\tag{4.24}
$$

From Eqs. (4.21), (4.22) and (4.24), we get

$$
\begin{aligned}
s^*_{jl} &= \max\left(0, \, V_{jl} - \frac{(1 + \lambda)V_{jl} - (\mathbf{X}^\mathsf{T}\mathbf{U})_{jl}}{1 + \lambda}\right) - V_{jl}, \\
&= \max\left(0, \, \frac{(\mathbf{X}^\mathsf{T}\mathbf{U})_{jl}}{1 + \lambda}\right) - V_{jl}.
\end{aligned}
\tag{4.25}
$$

The detailed update of $\mathbf{V}$ is given in Algorithm 3.

Algorithm 3 Closed-form update for $\mathbf{V}$

1: Given $\mathbf{U}_{k+1}, \mathbf{V}_k, \mathbf{X}, k \geq 1$.
2: for each $i \in (1, n)$, $j \in (1, r)$
    Compute $s^*_{i,j}$ by (4.25).
    Set $V^{i,j}_{k+1} = s^*_{i,j} + V^{i,j}_k$.
    end
3: Output $\mathbf{V}_{k+1}$ in order to update $\mathbf{U}$ in next iteration.

4.2.5   Convergence Analysis

Proposition 4.1 *Let* $\{\mathbf{U}_k, \mathbf{V}_k\}$ *be an infinite sequence of iterates generated by Algorithm 1. Then every accumulation point of* $\{\mathbf{U}_k, \mathbf{V}_k\}$ *is a critical point of* f *over the space* $\mathcal{M}_r \times \mathbb{R}_+^{n \times r}$, *namely* $\lim_{k \to \infty} gradF(\mathbf{U}_k, \mathbf{V}_k) = \mathbf{0}$ *and* $\lim_{k \to \infty} \partial_{\mathbf{V}_k} F(\mathbf{U}_k, \mathbf{V}_k) = \mathbf{0}$, *where* $gradF(\mathbf{U}_k, \mathbf{V}_k) = gradf(\mathbf{U}_k)$.

Proof Note that $\mathcal{M}_r$ is a compact manifold. Moreover, $\{\mathbf{V}_k\}$ is bounded; otherwise $F(\mathbf{U}, \mathbf{V})$ will go to infinity. Without loss of generality, suppose $\mathbf{V} \in [0, L]^{n \times r}$, where $L > 0$ is a finite number. Now both $\{\mathbf{U}_k\}$ and $\{\mathbf{V}_k\}$ stay in a closed and bounded subset.

We complete the proof by contradiction. Without loss of generality, suppose

$$\lim_{k \to \infty} ||gradF(\mathbf{U}_k, \mathbf{V}_k)||_F + ||\partial_{\mathbf{V}_k} F(\mathbf{U}_k, \mathbf{V}_k)||_F \neq 0,$$

and then there exists an $\epsilon > 0$, and a subsequence in $\{(\mathbf{U}_k, \mathbf{V}_k)\}$ such that

$$||gradF(\mathbf{U}_k, \mathbf{V}_k)||_F + ||\partial_{\mathbf{V}} F(\mathbf{U}_k, \mathbf{V}_k)||_F \geq 2\epsilon > 0$$

for all k. Since $\mathcal{M}_r$ is closed and bounded, and $\mathbf{V}_k$ is constrained in $[0, L]^{n \times r}$, the subsequence $\{(\mathbf{U}_k, \mathbf{V}_k)\}_{k \in \Gamma}$ should have a limit point $(\mathbf{U}^*, \mathbf{V}^*)$ on $\mathcal{M}_r \times [0, L]^{n \times r}$, i.e. $\lim_{k \to \infty} F(\mathbf{U}_k, \mathbf{V}_k) = F(\mathbf{U}^*, \mathbf{V}^*)$. By the continuity of $gradF(\mathbf{U}, \mathbf{V})$ and $\partial_{\mathbf{V}} F(\mathbf{U}, \mathbf{V})$, it implies that either

$$\lim_{k \to \infty} ||gradF(\mathbf{U}_k, \mathbf{V}_k)||_F \geq \epsilon > 0$$

or

$$\lim_{k \to \infty} ||\partial_{\mathbf{V}} F(\mathbf{U}_k, \mathbf{V}_k)||_F \geq \epsilon > 0.$$

Without loss of generality, suppose $\lim_{k \to \infty} ||gradF(\mathbf{U}_k, \mathbf{V}_k)||_F \geq \epsilon > 0$. Based on (4.16), there exist a step size α such that $C_k = f(\mathbf{U}_k)$ and $f(\mathbf{U}_{k+1}) = F(\mathbf{U}_{k+1}, \mathbf{V}_k) \leq f(\mathbf{U}_k) + \delta\alpha\langle gradf(\mathbf{U}_k), \boldsymbol{\eta}_k \rangle$. Note that $F(\mathbf{U}_k, \mathbf{V}_k) = f(\mathbf{U}_k) > F(\mathbf{U}_{k+1}, \mathbf{V}_k)$, $\boldsymbol{\eta}_k$ is a descent direction (such as the steepest descent direction), and $\lim_{k \to \infty} ||gradF(\mathbf{U}_k, \mathbf{V}_k)||_F \geq \epsilon > 0$. Then $\forall k$, it follows that $|F(\mathbf{U}_k, \mathbf{V}_k) - F(\mathbf{U}_{k+1}, \mathbf{V}_{k+1})| \geq |F(\mathbf{U}_k, \mathbf{V}_k) - F(\mathbf{U}_{k+1}, \mathbf{V}_k)| \geq |\delta\alpha\langle gradf(\mathbf{U}_k), \boldsymbol{\eta}_k \rangle| = v > 0$, where v is some constant. This contradicts that $\{(\mathbf{U}_k, \mathbf{V}_k)\}$ has a limit point. In addition, we have similar result for $\mathbf{V}$. We therefore conclude that $\lim_{k \to \infty} gradF(\mathbf{U}_k, \mathbf{V}_k) = \mathbf{0}$ and $\lim_{k \to \infty} \partial_{\mathbf{V}} F(\mathbf{U}_k, \mathbf{V}_k) = \mathbf{0}$.

4.3   Experimental Evaluation

In this section, we use both synthetic and real-world data sets to show the performance of our proposed method, NRCG-ONMF. Our method is implemented in Matlab and the source codes are available per request. The source codes of all the comparing methods are either publicly available or obtained from the corresponding authors. We perform the experiments on a PC with 64-bit Windows 7 operation system, 8 GB RAM and 2.40 GHZ Intel i7-3630QM CPU.

4.3.1 Implementation Details

4.3.1.1 Initialization

In general, optimization methods on nonlinear manifolds are guaranteed to converge to a local solution. Therefore, a good initialization of $\mathbf{U}_0$ and $\mathbf{V}_0$ is important. In our case, $\mathbf{U}_0$ and $\mathbf{V}_0$ can be obtained by applying truncated Singular Value Decomposition (SVD) of rank r on $\mathbf{X}$, namely $[\bar{\mathbf{U}}, \bar{\mathbf{S}}, \bar{\mathbf{V}}] = \mathrm{svds}(\mathbf{X}, r)$, where $\bar{\mathbf{U}}$, $\bar{\mathbf{S}}$ and $\bar{\mathbf{V}}$ are the output of SVD. We then set $\mathbf{U}_0 = \bar{\mathbf{U}}$ and $\mathbf{V}_0 = \max(\bar{\mathbf{S}}\bar{\mathbf{V}}, 0)$.

4.3.1.2 Gradient Computation

Because $\mathbf{U}$ is imposed on orthogonality constraint, namely $\mathbf{U}^{\mathsf{T}}\mathbf{U} = \mathbf{I}_r$, the following equation has the same projection result on the tangent space $T_{\mathbf{U}}\mathcal{M}_r$ as Eq. (4.7).

$$\mathbf{G}_{\mathbf{U}} = (\mathbf{U}\mathbf{V}^{\mathsf{T}} - \mathbf{X})\mathbf{V} \tag{4.26}$$

So to reduce calculation time, when calculating $\mathrm{grad} f(\mathbf{U})$, we use Eq. (4.26) to replace $\mathbf{G}_{\mathbf{U}}$ in Eq. (4.6).

4.3.1.3 Objective Value Computation

Note that the original objective function is $\frac{1}{2}||\mathbf{X} - \mathbf{U}\mathbf{V}^{\mathsf{T}}||_F^2 + \frac{\lambda}{2}||\mathbf{U}\mathbf{V}^{\mathsf{T}}||_F^2$, which has $O(mnr + mn)$ complexity. However, we can rewrite it into the following form:

$$\frac{1}{2}||\mathbf{X}||_F^2 - trace(\mathbf{X}^{\mathsf{T}}\mathbf{U}, \mathbf{V}) + \frac{1+\lambda}{2}||\mathbf{V}||_F^2. \tag{4.27}$$

Note that $\frac{1}{2}||\mathbf{X}||_F^2$ is a constant, thus it can be pre-computed. Now, the calculation of Eq. (4.27) takes $O(mnr + mr)$. As generally $r \ll n$, we can reduce the computation cost using Eq. (4.27) to compute objective value.

4.3.2 Data Sets

We use synthetic data to compare the orthogonality and convergence of various ONMF algorithms. Real-world data sets are used for the comparison of clustering performance.

4.3.2.1   Synthetic Data

We generate a synthetic data set in order to control the noise level. Specifically, we select five base vectors $\mathbf{b}_j, j \in [1, 5]$ randomly from the unit hypercube in n dimensions (here $n = 100$). The selected base vectors are independent to each other. Then, we generate data points $\mathbf{x}_i = w_i \mathbf{b}_j + \epsilon \mathbf{v}_i, i \in [1, m]$ (here $m = 100$ is the number of data points generated for a testing matrix), where $\epsilon \in [10^{-2}, 1]$ is a parameter controlling the noise level, $w_i \in (0, 1)$ is the random weight on base vectors and satisfies the uniform distribution. $\mathbf{v}_i$ is the random vector satisfying the multivariate standard Gaussian distribution. Negative entries of $\mathbf{x}_i$ are set to zero. For each $\epsilon \in [10^{-2}, 1]$, we generate m data points and vertically concatenate them into a $m \times n$ matrix $\mathbf{X}$, where each row represents a data point.

4.3.2.2   Real-World Data Sets

We collect several publicly-available real-world data sets for our experimental studies. The datasets we choose are widely used in the related works thus easily to compare the performances between these works and our work. These data sets are described as follows:

COIL-20 [155]. It is an image data set of objects, which contains 1440 gray scale images of 20 different objects. In our experiment, each image is resized to 32×32 pixels.

The Yale Face Database [156]. This data set contains 165 gray scale images of 15 people. Each one has 11 images with different facial expression or configuration. We resize each image to 32×32 pixels.

CLUTO. It is a data set collected for text mining [157]. We choose one large subset (k1b) and one medium subset (wap) from CLITO to showcase the performance on text data with various sizes.

UCI-mfeat [158]. This data set consists of features of handwritten numerals extracted from a collection of Dutch utility maps. 200 patterns per class for a total of 2000 patterns have been digitized in binary images files.

Table 4.2 shows the statistics of each of the data sets.

Table 4.2 Data sets description

Data	Instance#	Dimension#	Class#	Type
COIL20	1440	1024	20	image
Yale	165	1024	15	image
CLUTO-k1b	2340	21,839	6	text
CLUTO-wap	1560	8460	20	text
UCI-mfeat-fac	2000	216	10	text
UCI-mfeat-fou	2000	76	10	text
UCI-mfeat-pix	2000	240	10	text
UCI-mfeat-zer	2000	47	10	text

4.3.3 Metrics

4.3.3.1 ONMF Metrics

We evaluate the various ONMFs in terms of three metrics, namely relative *Frobenius* approximation error (relative error for short thereafter), orthogonality, computation time. They are defined as follows:

- Relative error: we define the relative error as in [144]:

$$RelFerr = \frac{||\mathbf{X} - \mathbf{U}\mathbf{V}^T||_F^2}{||\mathbf{X}||_F^2} \tag{4.28}$$

- Orthogonality: we leverage the orthogonality measurement in [143]: $||\mathbf{U}^T\mathbf{U} - \mathbf{I}||$.
- Computation time: we record the time an algorithm takes to converge or reach the maximum iteration.

4.3.3.2 Clustering Metrics

To evaluate the clustering performance, we adopt three metrics, namely Clustering Accuracy (CA), Normalized Mutual Information (NMI) and Purity metrics from [136, 141, 159, 160]. The definitions of the three metrics are given as follows:

- **CA** is defined as follows:

$$CA = \frac{\sum\limits_{i=1}^{n} \delta(map(r_i), l_i)}{n} \tag{4.29}$$

where r_i is the computed cluster label and l_i is the true cluster label. $\delta(x, y) = 1$ if $x = y$, otherwise, $\delta(x, y) = 0$. $map(r_i)$ is a mapping function that matches the computed label to the best true label. We adopt the Kuhn-Munkres [161] algorithm for the mapping. A larger CA value indicates a better clustering result.

- **NMI** is defined as follows:

$$NMI = \frac{I(R, L)}{(H(R) + H(L))/2} \tag{4.30}$$

where R denotes the set of true cluster labels and L is a set of computed lables from the evaluated algorithm. $I(R, L)$ is the mutual information (see [162] for the definition) between R and L. $H(.)$ is the entropy function. A larger NMI value indicates a better clustering result.

- **Purity** measures the extent to which each cluster contained data points from primarily one cluster. The purity of a clustering algorithm is obtained by the weighted sum of individual cluster purity values [160], given as:

$$Purity = \frac{1}{n} \sum_{i=1}^{k} max_j(n_i^j) \qquad (4.31)$$

where n_i^j is the number of the i-th input cluster that is assigned to the j-th cluster, k is the number of clusters and n is the total number of the data points. A larger purity value indicates a better clustering solution.

4.3.4   Results

In the experiments, we use the stopping criteria suggested in [149]: when $|1 - \frac{\sqrt{2*currentObj}}{\sqrt{2*preciousObj}}| < 1e-5$, the iteration stops. The $currentObj$ is current objective value and $preciousObj$ is the objective value from last iteration.

4.3.4.1   Convergence Comparison

We evaluate the convergence of NRCG-ONMF with three β solutions and ONPMF on synthetic data without noises ($\epsilon = 0$). The other two β solutions besides HS (Eq. (4.12)) are as follows:

Polak-Ribiére (PR) rule [148] is:

$$\beta_k = \frac{\langle \mathrm{grad} f(\mathbf{U}_k), \mathrm{grad} f(\mathbf{U}_k)\rangle - \langle \mathrm{grad} f(\mathbf{U}_k), \mathrm{grad} f(\mathbf{U}_{k-1})\rangle}{\langle \mathrm{grad} f(\mathbf{U}_{k-1}), \mathrm{grad} f(\mathbf{U}_{k-1})\rangle}.$$

and the Fletcher-Reeves (FR) rule [163] is:

$$\beta_k = \frac{\langle \mathrm{grad} f(\mathbf{U}_k), \mathrm{grad} f(\mathbf{U}_k)\rangle}{\langle \mathrm{grad} f(\mathbf{U}_{k-1}), \mathrm{grad} f(\mathbf{U}_{k-1})\rangle}.$$

We choose ONPMF because it also considers *Stiefel* manifold when preserving orthogonality for **U**. Three randomly generated matrices with different dimensions and ranks are used in our experiments and we set the maximum iteration number to 50 in order to show the different performance clearly. Figure 4.1 depicts the results. NRCG-ONMF algorithms are shown non-monotone because BB step does not necessarily decrease the objective value at every iteration, but this issue has been solved in our adopted non-monotone line search method in [153]. For the small and medium sized matrices (100×100 and 5000×5000), all variants of NRCG-ONMF (i.e., with different β) can converge within 50 iterations (Fig. 4.1a, c and e). The convergence time on small matrix is less than 0.07 s and the time on medium matrix is less than 30 s (Fig. 4.1b, d and f). For small matrix, NRCG-ONMF(HS) converges less quickly than NRCG-ONMF(FR) and NRCG-ONMF(PR), but is more quickly

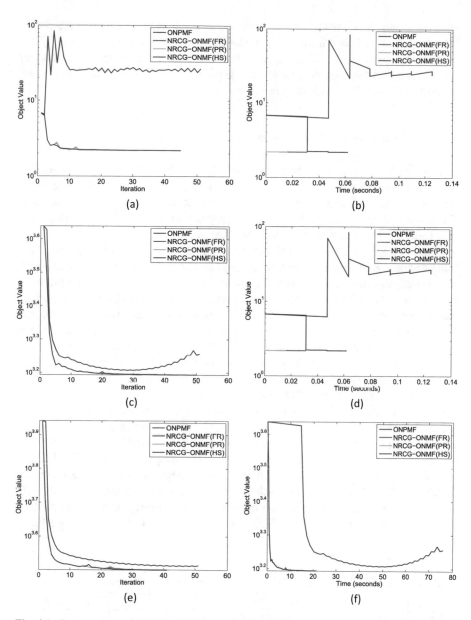

Fig. 4.1 Convergence of NRCG-ONMFs and ONPMF on synthetic data sets of different sizes/ranks. Three NRCG-ONMF variants with different β choices are compared. The matrices are generated randomly, and the maximum iteration number is set to 50. (**a**) 100×100, $r = 5$. (**b**) 100×100, $r = 5$. (**c**) 5000×5000, $r = 50$. (**d**) 5000×5000, $r = 50$. (**e**) 5000×5000, $r = 100$. (**f**) 5000×5000, $r = 100$

than these two algorithms when performed on medium matrix. Thus we only choose
NRCG-ONMF(HS) in the following evaluations. ONPMF cannot converge within
50 iterations (more than 500 iterations when raising the iteration limit). For medium
matrix with rank 50, it has a up trend on objective values. NRCG-ONMF algorithms
outperform ONPMF in all testing matrices.

4.3.4.2 Orthogonality Comparison

In this section, we compare several state-of-the-art ONMF works with our method
NRCG-ONMF(HS). We also show the comparison of their performance when
introducing noises. It is worth noting that the parameters of each comparing methods
are set as the values suggested in the corresponding papers. The descriptions of
compared ONMF works are shown in the Table 4.3.

Table 4.4 gives the average value of comparison results based on 100 tries. The
values of bold font denote the best performance values. The iteration number equals
to 500 indicates the algorithm cannot converge within 500 iteration. From this table
we can see that NRCG-ONMF(HS) achieves the lowest values in three metrics
except the iteration number. SPANONMF is the second fastest algorithm. However
it fails to maintain the orthogonality constraint. ONPMF is the only one that has
comparable performance with NRCG-ONMF(HS) in terms of keeping orthogonality. However it has one order higher relative error than NRCG-ONMF(HS), as well
as higher computation time and number of iterations.

We further compare NRCG-ONMF(HS), SPANONMF and ONPMF when the
noises are introduced. We run each algorithm for 100 times and record the average
value. We choose ONMF because it maintains the only comparable performance
on preserving orthogonality with our method NRCG-ONMF(HS). SPANONMF
is chosen as it is the second best in terms of computation time (Table 4.4).
Figure 4.2 shows that the relative error increases when the noise level increases.
NRCG-ONMF(HS) continues to achieve the lowest relative error. SPANONMF
shows smoothly increased relative error that is consistently higher than NRCG-
ONMF(HS), while the value of ONPMF fluctuates at the beginning and gradually
becomes smooth.

Table 4.3 Description of orthogonal NMF algorithms discussed in this work

Algorithms	Object function	Orthogonal part
ONMFEU [138]	$min\|\|\mathbf{X} - \mathbf{FG}^T\|\|_F^2$	$\mathbf{F}$
ONMFA [143]	$min\|\|\mathbf{X} - \mathbf{AS}\|\|_F^2$	$\mathbf{A}$
NMFOS_ED [141]	$min\|\|\mathbf{V} - \mathbf{WH}\|\|_F^2 + \lambda\|\|\mathbf{W}^T\mathbf{W} - I\|\|_F^2$	$\mathbf{W}$
ONPMF [135]	$min\|\|\mathbf{M} - \mathbf{UV}\|\|_F^2$	$\mathbf{V}$
EMONMF [135]	$min\|\|\mathbf{M} - \mathbf{UV}\|\|_F^2$	$\mathbf{V}$
SPANONMF [144]	$min\|\|\mathbf{M} - \mathbf{WH}^T\|\|_F^2 + \epsilon\|\|\mathbf{M}\|\|_F^2$	$\mathbf{W}$
NRCG-ONMF(HS) (this work)	$min\|\|\mathbf{X} - \mathbf{UV}^T\|\|_F^2 + \lambda\|\|\mathbf{UV}^T\|\|_F^2$	$\mathbf{U}$

Table 4.4 Performance comparison of various ONMF methods

Algorithms	RelFerr	Orthogonality	Time (s)	#Iter
ONMFEU	231.1850	12,906.1000	0.2362	500
ONMFA	0.0989	2.1696	0.1381	333
NMFOS_ED	0.2323	0.4600	0.2293	500
ONPMF	0.3451	5.5267e−15	0.7849	500
EMONMF	0.0836	49.0612	0.2868	**10**
SPANONMF	0.1179	1.7320	0.0893	500
NRCG-ONMF(HS)	**0.0236**	**1.0438e−15**	**0.0747**	55

RelFerr represents relative Frobenius approximation error. Orthogonality denotes the value of $||U^T U − I||_F$. Time is the computation time for 500 iterations or until converged. #Iter is the actual iteration number. A lower value of all metrics indicates a better performance. Bold font denotes the best performance. The values are the average of 100 tries on $100 \times 100, rank = 5$ randomly generated matrices

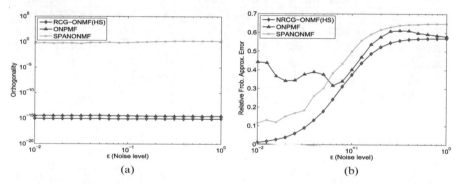

Fig. 4.2 Performance on synthetic data with noises. The values are the average of 100 tries on $100 \times 100, rank = 5$ randomly generated matrices. (**a**) Orthogonality. (**b**) Relative error

4.3.4.3 Clustering Performance Comparison

We compare various ONMF works on the clustering application. As EMONMF and ONMFEU performs the worst in keeping orthogonality and NMFOS_ED does not perform well in three metrics, we omit them in this evaluation. We also compare the well-known clustering algorithm k-means and some of the very recent matrix factorization works that aim to do clustering, namely CAPNMF [136], CVXNMF [145] and NMF_GCD [154].

We report the performance results in Tables 4.5, 4.6, and 4.7 for CA, NMI and Purity respectively with deviations after '±'. Values in bold denote the best performance. For most of the data sets, our algorithm NRCG-ONMF achieves the best in terms of accuracy, mutual information and purity. ONPMF and EMONMF also perform well in some data sets. For example, ONPMF gives the highest accuracy in the mfeat-fou data set and the highest purity in the COIL20 data set. EMONMF achieves the best NMI and purity for the Yale data set. k-means also

Table 4.5 Clustering performance comparison in terms of CA on different data sets

Data set	K-Means	CAPNMF	CVXNMF	NMF_GCD
COIL20	0.5364±0.0422	0.6942±0.0295	0.6601±0.0360	0.6390±0.0195
Yale	0.4353±0.0138	0.4532±0.0301	0.4345±0.0198	0.3830±0.0325
CLUTO-k1b	0.7579±0.0072	0.5940±0.0012	0.6460±0.0021	0.7015±0.0474
CLUTO-wap	0.3638±0.0420	0.2186±0.0124	0.3631±0.0138	0.3446±0.0245
UCI-mfeat-fac	0.6213±0.0552	0.6005±0.0302	0.5308±0.0402	0.4766±0.0479
UCI-mfeat-fou	0.623±0.0694	0.6013±0.0303	0.6894±0.0313	0.6306±0.0353
UCI-mfeat-pix	**0.7582±0.0572**	0.6578±0.0245	0.6673±0.0289	0.7071±0.0489
UCI-mfeat-zer	**0.5181±0.0160**	0.4705±0.0235	0.5058±0.0280	0.4025±0.0688

Data set	ONPMF	EMONMF	SPANONMF	NRCG-ONMF
COIL20	0.6736±0.0345	0.6260±0.0358	0.6011±0.0317	**0.6982±0.0306**
Yale	0.4321±0.0284	0.4333±0.0285	0.3770±0.0117	**0.4588±0.0257**
CLUTO-k1b	0.6665±0.0322	0.5726±0.1204	0.6633±0.1025	**0.7590±0.0492**
CLUTO-wap	0.3623±0.0098	0.3687±0.0231	0.2804±0.0114	**0.4090±0.0221**
UCI-mfeat-fac	0.5449±0.0549	0.5386±0.0621	0.2338±0.0003	**0.6472±0.0068**
UCI-mfeat-fou	**0.7201±0.0024**	0.6083±0.0429	0.5560±0.0521	0.7197±0.0006
UCI-mfeat-pix	0.6949±0.0672	0.5845±0.0976	0.5323±0.0857	0.7372±0.0624
UCI-mfeat-zer	0.5094±0.0045	0.4788±0.0566	0.3760±0.0807	0.5157±0.0182

Table 4.6 Clustering performance comparison in terms of NMI on different data sets

Data set	K-Means	CAPNMF	CVXNMF	NMF_GCD
COIL20	0.7186±0.0092	0.7556±0.0195	0.7696±0.0172	0.7425±0.0058
Yale	0.5011±0.0229	0.4935±0.0190	0.4973±0.0144	0.4242±0.0301
CLUTO-k1b	0.3732±0.0153	0.3027±0.0123	0.3727±0.0039	0.4347±0.0202
CLUTO-wap	0.3674±0.0277	0.3120±0.0106	0.4238±0.0154	0.3829±0.0141
UCI-mfeat-fac	0.6145±0.0244	0.5045±0.0356	0.5051±0.0157	0.4462±0.0698
UCI-mfeat-fou	0.6315±0.0318	0.5045±0.0308	0.6600±0.0081	0.5977±0.0105
UCI-mfeat-pix	0.6336±0.0301	0.5065±0.0284	0.6675±0.0172	0.6542±0.0187
UCI-mfeat-zer	0.4808±0.0097	0.3043±0.0403	0.4762±0.0151	0.3512±0.0589

Data set	ONPMF	EMONMF	SPANONMF	NRCG-ONMF
COIL20	0.7754±0.0242	0.7452±0.0224	0.6997±0.0194	**0.7826±0.0104**
Yale	0.4906±0.0173	**0.5050±0.0307**	0.4297±0.0233	0.5028±0.0231
CLUTO-k1b	0.4263±0.0123	0.2302±0.1350	0.3408±0.1223	**0.4638±0.0080**
CLUTO-wap	0.3946±0.0014	0.3727±0.0352	0.3927±0.0141	**0.4460±0.0142**
UCI-mfeat-fac	0.5192±0.0242	0.5134±0.0344	0.1936±0.0005	**0.6209±0.0061**
UCI-mfeat-fou	0.6672±0.0024	0.5704±0.0355	0.5339±0.0603	**0.6734±0.0018**
UCI-mfeat-pix	0.6568±0.0263	0.5492±0.0993	0.5128±0.0621	**0.6954±0.0242**
UCI-mfeat-zer	0.4728±0.0013	0.3992±0.0402	0.3008±0.0976	**0.4886±0.0028**

Table 4.7 Clustering performance comparison in terms of Purity on different data sets

Data set	K-Means	CAPNMF	CVXNMF	NMF_GCD
COIL20	0.6722±0.0490	0.6320±0.0115	0.6967±0.0105	0.6796±0.0137
Yale	0.4339±0.0447	0.4215±0.0205	0.4400±0.0175	0.3770±0.0224
CLUTO-k1b	0.7937±0.0108	0.5957±0.0102	0.7792±0.0004	0.7938±0.0121
CLUTO-wap	0.4901±0.0186	0.4256±0.0175	0.5004±0.0165	0.4827±0.0231
UCI-mfeat-fac	0.6610±0.0601	0.5090±0.0754	0.5125±0.0125	0.5081±0.0827
UCI-mfeat-fou	0.6844±0.0433	0.6039±0.0195	0.6875±0.0159	0.6541±0.0261
UCI-mfeat-pix	0.7130±0.0647	0.6056±0.0246	0.7104±0.0208	0.7312±0.0339
UCI-mfeat-zer	0.5499±0.0237	0.6045±0.0308	0.5497±0.0164	0.4350±0.0602

Data set	ONPMF	EMONMF	SPANONMF	NRCG-ONMF
COIL20	**0.7162+0.0279**	0.6614±0.0325	0.6247±0.0433	0.7122±0.0212
Yale	0.4327±0.0289	**0.4558±0.0367**	0.4218±0.0406	0.4364±0.0283
CLUTO-k1b	0.7935±0.0092	0.6776±0.0905	0.7238±0.0740	**0.8021±0.0106**
CLUTO-wap	0.5354±0.0108	0.4945±0.0242	0.4794±0.0210	**0.5413±0.0099**
UCI-mfeat-fac	0.6871±0.0018	0.5791±0.0537	0.3002±0.0354	**0.6883±0.0092**
UCI-mfeat-fou	0.7016±0.0157	0.6186±0.0333	0.5917±0.0383	**0.7202±0.0013**
UCI-mfeat-pix	0.7056±0.0268	0.6024±0.0951	0.5910±0.0297	**0.7539±0.0505**
UCI-mfeat-zer	0.5491±0.0019	0.4936±0.0489	0.4024±0.0363	**0.5506±0.0121**

shows good performance on some data sets. For example, it achieves the highest accuracy for UCI-mfeat-pix and UCI-mfeat-zer data sets.

4.4 Related Works

In this section, we review some of the representative efforts, as well as the most recent solutions of the ONMF problem. We also briefly overview the works that introduce manifold into the NMF solutions.

Ding et al. [138] are the first to explicitly propose the concept of ONMF. They impose the orthogonality constraint on factor matrix by considering Lagrangian multiplier which can be solved as an unconstrained optimization problem. Thus they apply the standard multiplicative update rules on each of the factor matrices. Choi [143] simplifies Ding's algorithm by turning the orthogonality constraint into *Stiefel* manifold. They directly use the gradient in *Stiefel* manifold in the multiplicative update of the orthogonal matrix. The Euclidean distance based ONMF method proposed in [141] also embeds Lagrangian multiplier in the solution. Moreover, they consider gradient descent in both the orthogonal subspace and the original space. However, these approaches produce heavy computation overhead as the Lagrange multiplier is a symmetrical matrix with many parameters.

Pompili et al. [135] propose two ONMF solutions, one of which is an EM-like ONMF algorithm based on the equivalence of ONMF and spherical k-means they proved. Very recently, Asteris et al. [144] propose an ONMF solution that relies on a novel approximation to the Non-negative Principle Component Analysis (NNPCA) problem, which jointly optimizes multiple orthogonal non-negative components and provably achieves an object value close to be optimal. However above works can only attain orthogonality to a limited extent. One exception is the method proposed by Pompili et al. in [135]. This method updates the orthogonal constrained matrix using projection step, which projects the matrix onto a feasible set of $St(k, n)$ of orthogonal matrices via a projection gradient method. The step size is chosen according to a backtracking line search using the Armijo rule. Our work shares similar idea with this method, but exploits more efficient nonlinear conjugate search algorithm with non-monotone step sizes.

Many research efforts introduce the manifold regularization into the NMF solutions [164–168]. Although they use the manifold concept in the matrix approximation problem, these approaches differ from our solution in that they identify the geometrical structure of the original data space by incorporating the geometrical structure into objective function (as the regularizer). Our solution considers the parameter constraint on factor matrix that is equivalent to the manifold definition.

4.5   Summary

In this chapter, we have proposed a NRCG-ONMF method which alternatively updates the orthogonal factor $\mathbf{U}$ by doing nonlinear search on *Stiefel* manifold, and updates the nonnegative factor $\mathbf{V}$ in a coordinate manner with closed form solutions. The convergence of NRCG-ONMF has been analyzed. Our approach sheds lights on an promising way to efficiently perform ONMF and shows great potential to handle large scale problems. We evaluate the proposed method on clustering tasks. Extensive experiments on both synthetic and real-world data sets demonstrate that the proposed NRCG-ONMF method outperforms other ONMF methods in terms of the effectiveness on preservation of orthogonality, optimization efficiency and clustering performance.

In the next chapter, we will present our work for knowledge extraction from unstructured data.

Chapter 5
Knowledge Extraction from Unstructured Data on the Web

In this chapter, we leverage Natural Language Processing (NLP) techniques, such as word embedding and topic model, to build a knowledge extraction system for source code repositories. The system automatically extracts topics/concepts from source code of software projects, aiming at facilitating the comprehension of the software projects. In addition to using topic model for topic extraction, we design a new algorithm that incorporates the word embedding learned from a large corpus of source code. To identify the most contributory features from source code for the topic extraction, we develop an automated feature selection process that select features for distinct projects independently. We evaluate our system using open source code of projects from Github.[1] With the proposed topic extraction and feature selection algorithms, the proposed system can effectively obtain topics from source code. This chapter is based on our research reported in [169].

5.1 Design Overview

Source code repositories and open source systems are becoming popular resources for software development because developers are able to reuse the shared source code of existing systems to build their own software [170]. The sharing virtue helps accelerating the development process. However, many such systems have poor or no documentation. For small software systems, one can understand its functions by reading and navigating source code. But it is time-consuming to browse all the source code files for large systems. Moreover, it requires non-trivial effort to identify useful ones from a large corpus of available systems.

Many works apply text summarization techniques to summarize source code [171–173] or parts of source code [174–177] aiming to assist the comprehension.

---

[1] https://github.com/.

© Springer International Publishing AG, part of Springer Nature 2018
W. E. Zhang, Q. Z. Sheng, *Managing Data From Knowledge Bases: Querying and Extraction*, https://doi.org/10.1007/978-3-319-94935-2_5

In this way, the source code can be comprehended by reading generated summaries. These approaches follow a common process that first extracts top-N keywords from source code and then builds summaries using these keywords. Thus, the quality of summarization depends on the extracted keywords, which represent the topics (we use *keywords* and *topics* interchangeably in this chapter hereafter). However, these works fail to discover latent topics by only considering term frequencies [172, 178].

Word embedding is a group of feature learning techniques in Natural Language Processing (NLP) that maps the words into vectors which are regarded as the underlying representation of the original words. Neural network is one of the methods for learning word embedding. It provides simple and efficient neural architecture to learn word vectors and has shown desirable performance in NLP tasks such as measuring word-to-word relationships [179]. The learned representations of two words reflect the latent relationship of the words, which can be used to discover the latent topics of documents. However, these techniques haven't been adopted in mining source code topics. In our system, we propose the first attempt to apply word embedding techniques to mining source code topics. For this purpose, we propose an extraction algorithm, Embedded Topic Extraction (EmbTE), that leverages the underlying relationships among terms from source code files via learned term vectors. Specifically, we first obtain vectors for terms in source code files using a neural embedding method proposed in [180]. Thus each document is mapped to a group of term vectors. Then we build a k-dimensional tree (k-d tree) [181] for term vectors belonging to one document. The centroid of the term vectors in this document is computed. Top-N nearest neighbours to the centroid are obtained by searching the k-d tree and are regarded as the top keywords (i.e., topics). The knowledge obtained from the topics can be used not only in source code summarization, but also in many more software engineer applications, such as software categorization, software clustering and software recommendation.

For comparison, we examine the effectiveness of the most popular topic model Latent Dirichlet Allocation (LDA) [182] in mining topics from source codes. LDA is a topic model that considers the term distribution in documents. It has shown effectiveness in finding latent topics from natural language text corpus. But limited works apply LDA to extract topics from source code. A classic dimension reduction method Non-negative Matrix Factorization (NMF) [130] is also evaluated for our topic extraction task. NMF factorizes a non-negative matrix to two matrices that are in lower dimension space. NMF can be adopted to topic extraction with the document-term matrix as the original matrix. The dimension of the two factored matrices is the number of topics. One factored matrix is document-topic matrix and the other one is topic-term matrix.

The second problem we address in this chapter is the evaluation metric of source code topic extraction methods. Existing works involve human expertise to perform the evaluation. Except for the works use pure human judgement [172, 174], other works compare the extracted keywords with human picked ones by adopting distance measurements on top-N ranked lists [178]. However, all these methods highly depend on human efforts, which have threats to validity. Instead, we adopt the evaluation metric for topic models in our work. Specifically, we use coherence measurement, which measures the coherence among extracted topics. A list of extracted topics with high coherence are considered as good extraction. An algorithm that combines the indirect cosine measure with Normalized Pointwise Mutual Information (NPMI) and the boolean sliding window [183] is used.

The third problem we consider in this chapter is to identify the most contributory types of terms for source code topic extraction. Existing works consider comments [171, 178], method names [173], class names and identifier names [172] in source code topic extraction. But limited explanation is given on why these types of terms are chosen and how the terms affect the source code topic extraction performance. Therefore, we develop an automated process that heuristically select the types of terms that increase the coherence of extracted topics. The ones who reduce the coherence will be ignored. The proposed selection approach draws ideas from the wrapper family in feature selection algorithms [184] in the machine learning community.

In a nutshell, the main contributions of our work are summarized as follows:

- We propose a source code topic extraction method EmbTE by leveraging the word embedding techniques. We compare EmbTE with LDA and NMF on extraction topics from source code. The coherence measurement is used as the performance metric which has not been adopted in existing works on mining source code.
- We develop a selection process that automatically identifies the most contributory terms of source code for topic extraction. This automated process can also be applied to other software engineering applications (e.g., link recovery, bug location and component categorization) by modifying the performance assessment metric.
- We evaluate our methods on Github Java projects. Although only Java projects are considered in this work, our approach is language-independent.

The remainder of this chapter is structured as follows. We describe our extraction methods in detail in Sect. 5.2. Section 5.3 reports the experimental results. Existing related research efforts are discussed in Sect. 5.4.

5.2 Source Code Topics Extraction via Topic Model and Words Embedding

To extract topics from source code, we propose a system that contains three main steps to extract topics from source code of projects. Figure 5.1 depicts these three main steps. We first pre-process the source code files (Sect. 5.2.1), then apply topic extraction methods (Sect. 5.2.2). We use coherence measurement as metric to evaluate topic extraction performance (Sect. 5.2.3). Finally, we discuss the proposed automated terms selection algorithm that selects the most contributory terms (Sect. 5.2.4) for the topic extraction.

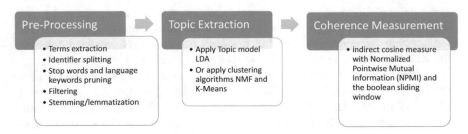

Fig. 5.1 Three main steps in system of extracting topics from source code

5.2.1   Data Pre-processing

The pre-processing step focuses on extracting feature terms from source code which are used as input to topic extraction methods. Unlike text document, the source code consider much programming language-specific terms, so our goal is to prune the invalid and meaningless ones and construct clean input to further steps. The details of pre-processing are as follows:

- *Feature terms extraction*. We extract feature terms such as method names, class names from source code. Each type of feature terms form a document. For example, all the method names of one project are in one document.
- *Identifier splitting*. Unlike in natural language text, where each word can be found in the dictionary, source code identifier names are generally a combination of several words and delimiters following certain naming convention, e.g., "loanInterests". We split this kind of identifiers into a set of terms, e.g., "loanInterests" to "loan" and "Interests".
- *Stop words and language keywords pruning*. Programming language keywords such as "implements" in Java do not contribute to comprehend the projects. Hence we prune the language keywords. We also remove the stop words.
- *Stemming and lemmatization*. Like natural language text, terms are used in various forms such as singular and plural, comparatives and superlatives. In our analysis, we do not consider them as different terms. Thus we unify them by using stemming and lemmatization techniques.
- *Filtering*. Keywords such as "set" are very generic and not contributory to the overall comprehension. Thus we filter out such words in our analysis.

5.2.2   Topic Extraction

5.2.2.1   Topic Extraction via Words Embedding

The embedding methods [179, 180] proposed to reveal the semantic information embedded in documents, topics and terms. These methods represent words by learning essential concepts and representations and show effectiveness in many NLP and machine learning tasks. Mikolov et al. [180] propose an efficient method using neural networks to learn distributed representation of words from large data sets. The method includes both Continuous Bag of Words (CBOW) and Skip-gram models. Given a word sequence $S = \{w_1, \ldots, w_q\}$, the objective functions for learning based on CBOW and Skip-gram are as follows:

$$F_{CBOW}(S) = \frac{1}{q} \sum_{i=1}^{q} \log p(w_i|w_c), \qquad (5.1)$$

$$F_{Skip-gram}(S) = \frac{1}{q} \sum_{i=1}^{q} \sum_{-l \leqslant \delta \leqslant l, \delta \neq 0, j=i+\delta} \log p(w_j|w_i), \qquad (5.2)$$

where w_c in Eq. (5.1) denotes the context of the word w_i and δ in Eq. (5.2) is the window size of context. Both equations aim to maximize the corresponding log-likelihoods.

We adopt the learned vector representation of words in EmbTE. Figure 5.2 depicts how EbmTE works. We first extract terms from source code files of systems and construct one document for each system. The document is pre-processed according to Sect. 5.2.1. Then EmbTE learns words vectors from a corpus of documents $\{d_1, \ldots, d_n\}$, which represents a corpus of systems that need to extract topics. Each document d_i can be represented by a group of vectors $\{\mathbf{w}_1, \mathbf{w}_2, \ldots, \mathbf{w}_m\}$, where m is the number of words in d_i. EmbTE computes the centroid $\mathbf{w}_{ct}$ of $\{\mathbf{w}_1, \mathbf{w}_2, \ldots, \mathbf{w}_m\}$ using k-medoids clustering with only one cluster and cosine distance as the distance metric. $\mathbf{w}_{ct}$ is considered as the key topic of d_i. Then EmbTE searches $\mathbf{w}_{ct}$'s N nearest neighbors and considers them as the top-N keywords/topics of d_i. If $\mathbf{w}_{ct}$ represents a word in d_i, then only N-1 nearest neighbors are returned.

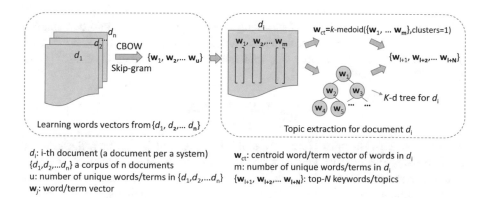

d_i: i-th document (a document per a system) $\mathbf{w}_{ct}$: centroid word/term vector of words in d_i
$\{d_1, d_2, \ldots d_n\}$ a corpus of n documents   m: number of unique words/terms in d_i
u: number of unique words/terms in $\{d_1, d_2, \ldots d_n\}$ $\{\mathbf{w}_{l+1}, \mathbf{w}_{l+2}, \ldots \mathbf{w}_{l+N}\}$: top-$N$ keywords/topics
$\mathbf{w}_j$: word/term vector

Fig. 5.2 Topic extraction via words embedding (EmbTE)

5.2.2.2 Topic Extraction via Latent Dirichlet Allocation

LDA is a widely applied topic model which has successfully extract topics from natural language texts. We adopt this model to extract source code topics. For completeness, we briefly introduce how LDA extracts topics. Given a corpus of document $D = \{d_1, \ldots, d_n\}$ where each document $d_i, i = 1, \ldots n$ is a sequence of m words denoted by $d_i = \{w_1, \ldots, w_m\}, w_j \in W, j = 1, \ldots m$. W is the vocabulary set. Each document d_i can be modelled as a multinomial distribution $\theta^{(d_i)}$ over t topics and each topic $z_k, k = 1, \ldots t$ is modelled as a multinomial distribution $\phi^{(k)}$ over the set of words W. LDA assumes a prior Dirichlet distribution on θ thus allowing the estimation of ϕ without requiring the estimation of θ. LDA is based on the following document generation process [182]:

- Choose $N \sim Poission(\xi)$: Select the number of words m.
- $\theta \sim Dir(\alpha)$: Select θ from the Dirichlet distribution parameterized by α.
- For each $w_j \in W$ do

  - Choose a topic $z_k \sim Multinomial(\theta)$
  - Choose a word w_j from $p(w_j|z_k, \beta)$, a multinomial probability ϕ^{z_k}.

LDA assumes each document is a word mixture of multiple topics. Thus the topic extraction is a reverse process of document generation.

To apply LDA for source code topics extraction, we construct one document per system's source code. As mentioned previously, pre-processed terms of one system construct a document. We assume the document contains terms/words from different topics as in natural language documents. Then we apply LDA on multiple documents. We use a variational Bayes approximation [182] of the posterior distribution in LDA. We first train LDA model on these document and obtain the terms distribution. Then top keywords for each document are returned. Thus we obtain topics from each document at the same time.

5.2.2.3 Topic Extraction via Non-negative Matrix Factorization

NMF considers the problem of factorizing a non-negative matrix into two smaller non-negative matrices. The problem is formulated as: Given $\mathbf{X} \in \mathbb{R}^{n \times m}$ and a positive integer $r < min(n, m)$, find non-negative matrices $\mathbf{U} \in \mathbb{R}^{n \times r}$ and $\mathbf{V} \in \mathbb{R}^{r \times m}$ which satisfy:

$$\min_{\mathbf{U}, \mathbf{V}} \ (||\mathbf{X} - \mathbf{UV}||_F^2), \tag{5.3}$$

where F is the Frobenius Norm. By multiplying $\mathbf{U}$ and $\mathbf{V}$ together, the product approximates the original matrix $\mathbf{X}$. NMF aims to reveal the meaningful latent features that are hidden in the input matrix $\mathbf{X}$, where each entry can be considered as the combination of these latent features with different weights. It can perform dimension reduction and clustering simultaneously. Thus it has enjoyed much success in many areas such as text mining, image processing and recommendation systems.

To extract topics using NMF, the input matrix $\mathbf{X}$ is the document-term matrix, where each entry is the weight of each term in each document. We use Term Frequency/Inverse Document Frequency (TF/IDF) as the weight. The dimension r is the number of topics. $\mathbf{U}$ represents the association between documents and topics while the $\mathbf{V}$ indicates the association between topics and terms. By ordering the values in a column of $\mathbf{U}$ and selecting the top ones, we can get the topics of the document. By ordering the values of rows of $\mathbf{V}$, we get the top terms/keywords for a topic. Thus we are able to associate document with keywords, which are considered as topics.

5.2.3 The Coherence Measurement

We adopt coherence measurement for evaluating the topics extraction performance because it does not require human effort to judge the extracted topics. Coherence measurement measures the coherence within the keywords list. High coherence score indicates that the extracted key words are closely coherent which further indicates that the extraction of topics is good. Michael Röder et al. [183] empirically show that the combination of indirect cosine measure with NPMI and the boolean sliding window gives best coherence measurement through their extensive studies on natural language corpus. We apply this measurement to the source code. Given a list of keywords $\{w_i\}, i \leq K$, where K is the number of keywords, four dimensions are required to measure the coherence on $\{w_i\}$: the segmentations of $\{w_i\}$, the probability P, the confirmation measurement m and the aggregation method. The segmentation of $\{w_i\}$ is defined as :

$$S_{set}^{one} = \left\{ (W', W^*) \| W' = \{w_i\}; w_i \in W; W^* = W \right\}, \tag{5.4}$$

where W is the total word set. The probability method used is boolean sliding window, which determines word counts using a sliding window. Confirmation measure takes a single pair $S_i = (W'; W^*)$ of word sublists as well as the corresponding probabilities to compute how strong the conditioning word set W' supports W^*. Indirect confirmation is defined as:

$$m_{s_{cos}(m,\gamma)}^{*}(W', W^*) = s_{cos}(\mathbf{u}, \mathbf{v}),$$

$$= \frac{\sum_{i=1}^{|W|} u_i \cdot v_i}{||\mathbf{u}||_2 \cdot ||\mathbf{v}||_2}, \tag{5.5}$$

where vector **u** is generated by:

$$\mathbf{u} = \left\{ \sum_{w_i \in W'} m(w_i, w_j)^\gamma \right\}, \tag{5.6}$$

where m, which is the direct confirmation measure using NPMI, is:

$$m_{nmpi}(S_i) = \frac{log \dfrac{P(W', W^*) + \varepsilon}{P(W') * P(W^*)}}{-log(P(W', W^*) + \varepsilon)}. \tag{5.7}$$

Similarly, vector $\mathbf{v} = \mathbf{v}_{m,\gamma}(W^*)$ is calculated using Eq. (5.6) by replacing W' with W^*. Finally, all confirmations of all subset pairs S_i are aggregated to a single coherence score using the arithmetic mean.

5.2.4   Automated Terms Selection for Topic Extraction

We have discussed the three extraction methods in previous section. Except for the extraction method, the terms/words used for extraction also heavily impacts the topic extraction performance. Existing works give limited explanation on why they use terms like comments, method names and identifier names for topic extraction. To this end, we propose a term selection algorithm that automatically selects the most contributory terms. We divide the terms into several types, e.g., method names, method comments, and identifier names. Each type of terms are recorded in a separate document. When a type is selected, the content of selected document is merged with the document that represents the system. We select the types of terms based on the performance of topic extraction (the coherence measurement discussed in Sect. 5.2.3) and use this metric to guide a heuristic search. The algorithm performs a best-first search in the feature space. It starts with constructing the terms set using one type of terms. Here we start from method names. Then it obtains top-N keywords by applying one of the extraction models discussed from Sects. 5.2.2.1 to 5.2.2.2. After obtaining keywords, the algorithm measures the contribution of the new selected type of terms and decide if it will be kept/removed. If it improves the topic extraction performance, the type of terms is selected. Otherwise, it is removed from the terms set. The algorithm then moves forward to construct a larger terms set by including more types of terms. In our work, we form a new document by merging the new selected term document with the term set document. Then the selection process is repeated until all types of terms are traversed.

Algorithm 1 describes the selection process. After initialization (line 1–2), it forwardly chooses a type of terms and evaluates its impact on the topic extraction performance, then decides when to keep this type of terms or remove it (line 3–12). Finally the algorithm outputs the selected terms (line 13). The algorithm adopts idea

---

Algorithm 1 Topic extraction guided feature selection

 1: Given a set of terms W, evaluation metric *metric*
 2: Initialize *mval*
 3: FOR each type of feature $W_{sub_i} \subset W$
 4:    *selected* = *selected.add*(W_{sub_i}),
 5:    Obtain keywords $\{w_1, ...w_k\}$ by applying topic extraction algorithms,
 6:    *newval* = *metric*($\{w_1, ...w_k\}$),
 7:    IF *newval* better than *mval*
 8:       *mval* = *newval*,
 9:    ELSE
10:       *selected* = *selected.remove*(W_{sub_i}),
11:    End IF
12: End FOR
13: Output the *selected*.

---

from feature selection in machine learning community, but it differs with it. Because in feature selection, the metric evaluation is directly performed on the set of features, while in this algorithm, the metric evaluation is based on the topics extracted from the extraction method, and the selected terms are the input of extraction model. This is the reason why we do not apply feature selection algorithms in our work.

5.3 Experimental Evaluation

We perform the experiments on Java projects randomly chosen from the Java project corpus provided by [185]. This corpus consists 14,807 Java projects in Github. All the .java files are kept and other files are removed. We compare EmbTE, LDA and NMF on extracting topics from these source codes. Both CBOW and Skip-gram models in EmbTE are considered. Firstly, we examine the impact of different weighting scheme on LDA and NMF methods. Second, we compare the performance of the three methods on various types of terms in the source codes. Third, we train the models on different number of systems and see the difference when the number changes. The coherence is computed via Eq. (5.5).

5.3.1 Setup

We extract source code terms from Java projects using JavaParser.[2] Stop words are pruned using Snowball English stop words lists.[3] We leverage StanfordNLP[4]

---

[2]https://github.com/javaparser/javaparser.
[3]http://snowball.tartarus.org/algorithms/english/stop.txt.
[4]http://nlp.stanford.edu/.

for the stemming and lemmatization. For the vector representation of words, we use gensim[5] implementation of Word2Vec.[6] The implementation of coherence measurement adopts Palmetto[7] tool. The scikit-learn[8] tool is used to implement LDA/NMF and build K-d tree for EmbTE. Other functions are implemented by us using Java. We perform all the experiments on a PC with Intel Core i7 2.40 GHZ processor and 8 GB RAM. All the source codes are available per request.

5.3.2   Results

The performance given in Sect. 5.3.2.1 is based on the evaluation of source codes from 100 systems. Sections 5.3.2.2 and 5.3.2.3 report the performances on the source codes from a small size of 10 and a medium size of 100 systems.

5.3.2.1   Impact of Different Weighting Schemes

In this experiment, we compare the performances of LDA and NMF on TF/IDF and TF weighting schemes when adding different types of terms. The bars from left to right show the coherence when adding new types of terms. We start from *MeNM* (method names), then the second bar of each group is the result based on merging *MeParNM* (method parameter names) to *MeNM*. Thus, until the last bar, the result on *VarType*, all the types of terms are added. We can see the coherence fluctuates when adding new types of terms. If the coherence depicted by one bar is higher than the value of previous bar, it means this type of terms increases the coherence (i.e., contributory). If the value is lower than the one of previous bar, this type of terms decreases the coherence (i.e., not contributory). Both Fig. 5.3a and b show the same trends (i.e., increase or decrease the coherence) when adding new types of terms. We will discuss in more details of the impact on different types of terms in Sect. 5.3.2.2 as we compare more methods there. In Fig. 5.3, we can also find that the TF/IDF weight gives better performance than only using TF as weight because TF considers only the frequency of the term in a document, while the TF/IDF also considers the number of documents that contain the term (using a logarithmically scaled inverse fraction of the documents that contain the term). Thus TF/IDF will lower the weight of terms that with high frequency in a document and appears in many documents. When we mine topics from a document, the ones appear in many documents will not well represent the topic of this document. That is why the extraction performances on TF/IDF weights are better than on TF weights.

---

[5]https://radimrehurek.com/gensim/.
[6]https://code.google.com/archive/p/word2vec/.
[7]https://github.com/AKSW/Palmetto.
[8]http://scikit-learn.org/.

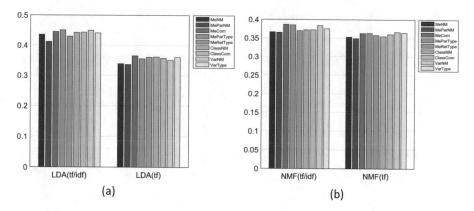

Fig. 5.3 Impact of different weights on topic extraction methods using LDA and NMF. The meaning of legends are as following: MeNM (method name), MeParNM (method parameter name), MeCom (method comments), MeParType (method parameter type), MeRetType (method return type), ClassNM (class name), ClassCom (class comments), VarNM (variable name) and VarType (variable type). (**a**) LDA. (**b**) NMF

5.3.2.2 Impact of Different Types of Terms

Figure 5.4 shows the topic extraction performance when adding terms. Figure 5.4a depicts the average value performance on 10 projects, while Fig. 5.4b depicts the average performance on 100 projects. The X-axis shows the results from adding different types of terms. Correspondingly, the Y-axis gives the coherence when the terms pointed by X-axis are added. For example, in Fig. 5.4b the coherence on *MeCom* (method comments) is higher than its previous coherence on *MeParNM*. It indicates that method names positively contribute to the topics coherence (i.e., improve performance). The coherence on *MeRetType* (method return types) is lower than its previous coherence which means method return types negatively impact the topic coherence (worsen performance). Figure 5.4a shows unstable trends among four considered models, but in Fig. 5.4b, the four models give similar performance on different types of terms. It indicates that the more documents are considered, the better topic extraction performance is achieved. When adding *MeNM* (method names), *MeCom* (method comments), *ClassNM* (class names) and *ClassCom* (class comments), the coherence increases, which means they are contributory to the topic extraction. The types of terms which decrease the coherence give negative impact on the topic extraction performance. We can also observe from the figures that EmbTE(CBOW) performs consistently the best and EmbTE(Skip-gram) has close performance with LDA. NMF shows the worst performance. EmbTE gives better performance than LDA is because that although the probability distribution obtained from LDA describes the statistical relationship among documents, topics and terms, sometimes the terms with higher probability can not represent topics well [186]. Instead, EmbTE considers semantic relationships between terms and is able to identify the central semantic of the documents. NMF performs worst is

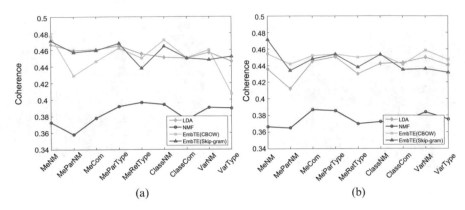

Fig. 5.4 Impact of different types of terms. The meaning of X-axis ticks are the same with legend meanings in Fig. 5.3. (**a**) On 10 projects. (**b**) On 100 projects

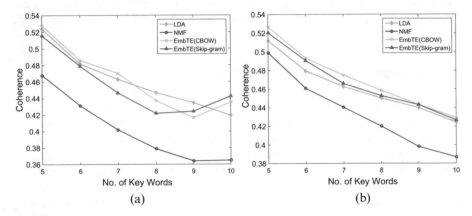

Fig. 5.5 Impact of number of keywords. (**a**) Impact of # keywords (10). (**b**) Impact of # kseywords (100)

because it sets fixed values for the probability of multinomial distribution of words and topics in documents, however, this is often unlikely in real documents. EmbTE with CBOW model performs better than with Skip-gram model. It indicates that the frequency of terms gives more impact than the context of terms in extracting topics from source codes.

5.3.2.3 Impact of Different Number of Keywords

According to the results in Sect. 5.3.2.2, we use the contributory types of terms, namely method name, method comments, class names and class comments from source codes in this experiment. Figure 5.5 gives the results when varying the number of keywords/topics. Both Fig. 5.5a and b show similar trends for four methods.

When the number of keywords is increasing, most methods produce lower coherent topics. EmbTE(CBOW) consistently performs the best while EmbTE(Skip-gram) and LDA performs the less best and NMF is the worst. The results show that more extracted topics will reduce the coherence of topics. This is because the coherence is normalized, when the new added topic is not 100% coherent with existing topics, the overall coherence will decrease.

5.4 Related Works

Much research effort has been devoted to tackle different tasks of source code mining. We discuss some of the works that are closely related to our work.

Source code summarization works apply text summarization techniques to generate summary for source code [171–173] or parts of source code [174]. Haiduc et al. [172] proposed an automated text summarization method using Vector Space Model (VSM). They extracted method comments, method names and class names as the terms and formed document-term matrix with TF/IDF weighing scheme. The terms with top-weighting are considered as the top keywords of the document. Finally, they used the keywords to construct the summary for the software system. The evaluation includes two parts: judgement from experienced developers and feedback from follow-up questionnaires. Rodeghero et al. [178] improved the work in [172] by introducing eye-tracking techniques. They also used method comments and VSM. Their method assigns weights according to the code position obtained from eye-tracking experiments. It uses *Kendall's* τ [187] distance to evaluate the keywords extraction by computing the distance between extracted keywords and the ones picked by experienced developers. Sridhara et al. contributed several works on generating comments and summaries for Java methods [174]. They designed heuristics to choose method statements and then transformed these statements to natural language comments and summaries. Their approach requires scanning the content of a method code.

Rama et al. [171] proposed to use topic model LDA to extract business concepts from source code. This work assumes a software system is a text corpus, in which the files are constructed using function names, comments and data structures. However, using this approach, each project requires an independent LDA model which is not applicable to processing large volume of projects at the same time. Many following works adopted LDA for solving various software engineer problems. For example, Asuncion et al. [188] applied LDA to enhance software traceability. They recorded traceability links during the software development process and learned a LDA model over artifacts. The learned model categorized artifacts according to their topics. Lukins et al. [189] used LDA to localize bugs in software systems. They learned topic model from source codes. Queries generated from bug reports were performed against the learned model to localize the code that might produce the bug.

5.5 Summary

In this chapter, we have developed a method EmbTE, for source code topic extraction, based on word embedding techniques. We also adopted LDA and NMF to extract topics from source code. The empirical comparisons show that EmbTE outperforms LDA and NMF in terms of providing more coherent topics. EmbTE with CBOW model performs better than Skip-gram model. We also identified the most contributory terms from source code via our proposed term selection algorithm. We found that the method name, method comments, class names and class comments are the most contributory term types.

After discussed knowledge extraction in this chapter, we will introduce our preliminary work towards building knowledge base from programming question answering communites, such as Stack Overflow.

Chapter 6
Building Knowledge Bases from Unstructured Data on the Web

In this chapter, we present the first attempt towards building a programming-centric knowledge base. We first propose a prototype to automatically extract knowledge from programming Question-Answering (QA) communities. Then we focus on identifying duplicate posts from programming community based question-answering (PCQA) websites based on the observation that PCQA websites contain duplicate questions despite the detailed posting guideline. Our method extract features that are able to capture semantic similarities between questions. These features are obtained by leveraging continuous word vectors from the deep learning literature, topic model features and phrases pairs that co-occur frequently in duplicate questions mined using machine translation systems. Experiments on a range of real world datasets demonstrate that our method produce a strong performance for duplicate detection. This chapter is based on our research reported in [169, 190].

6.1 Design Overview

Community based question answering (CQA) sites such as Quora,[1] Baidu Zhidao,[2] and Stack Exchange[3] have grown in popularity in the recent years. CQA sites are a promising alternative to traditional web search as users have queries that are often subjective, open-ended and require expert opinions. To cater the multitude of interests for its community, Stack Exchange has a set of sub-domains that focus on a particular subject or topic.

Stack Overflow, a programming community question answering site (PCQA), is a sub-domain in Stack Exchange created for programming related questions. It provides working code solutions to natural language queries and serves as a knowledge

---

[1] https://www.quora.com/.

[2] http://zhidao.baidu.com/.

[3] http://stackexchange.com/.

resource for software developers. In this chapter, we propose a system prototype that first targets the problem of extracting programming-centric knowledge from PCQA websites (e.g., Stack Overflow). It is the initial step of building a programming-centric knowledge base. We extract questions and answers for a post and form a triple formatted as ⟨*"question"*; *"answer"*; *"tag1", "tag2",...*⟩. The tags can be easily obtained and the answers can be obtained by choosing the accepted answers marked by the questioner. Then we extract information from question and rewrite them to the *"question"* component of knowledge triples. This can be done by using NLP techniques.

The key challenge lies in extracting information from questions because duplicate questions—questions that were previously created and answered—are a frequent occurrence even though users are reminded to search the forum before creating a new post. To reduce the number of duplicate questions, Stack Overflow encourages reputable users to manually mark duplicate questions. This approach is laborious, but more importantly, a large number of duplicate questions undetected (false negative). A high quality duplication detection system will considerably improve user experience: for inexperienced users creating a new question it can suggest a related post before posting; for experienced users it can suggest potential duplicate posts for manual verification.

Question duplication is a pervasive issue in CQA in general, and a number of studies have looked into related problems, including finding similar questions [191–194], and generating answers for questions [195, 196]. These works framed the task as a classification or prediction task, and relied on a number of extracted features to train a model. It is important to note, however, that features explored in these methods may not necessarily translate to PCQA as PCQA posts often contain source code from programming languages which are linguistically different to natural languages. There are few studies that explored question duplication for the PCQA domain [197, 198]. The work in [198] tackled duplicate question detection on Stack Overflow by generating features using title, description, topical and tag information to classify whether a question pair is a duplicate. The work in [197] improved on this methodology by extending features adopted from [199] which mined information from Twitter posts. Additionally, they conducted extensive analyses to learn why users create duplicate questions. In this chapter, we seek to improve upon the benchmark performance set by their system.

Our methodology follows the same approach from previous works by framing the duplication detection task as a supervised classification problem. Henceforth we refer to our system as PCQADup. Given a question pair, PCQADup generates three types of features. *Vector similarity*, the first of its features, represent questions as continuous vectors in a high-dimensional space. In the deep learning literature, word2vec [180] was proposed to learn word vectors/embeddings efficiently from large text corpora, and it has been shown to effectively capture semantics of words. doc2vec, an extension of word2vec, is developed to generate embeddings for any arbitrary word sequences [200]. Inspired by the success of these neural methods, we compute the doc2vec representation of the title and body content of a post, and measure similarity between a question pair based on vector cosine similarity measures. The second type of features is *topical similarity*, computed using a topic model for extracting themes from short texts. Similarity of a question pair is measured by computing the similarity of topical distribution between the pair. The last type of features is *association scores*. We first mine association pairs, i.e., pairs of phrases that co-occur frequently in known duplicate questions, by adopting a word alignment method developed in the machine translation literature. To generate the association score for a question pair, we train a perception that takes association pairs and lexical features as input. The idea of using associated phrases has been explored in knowledge base question answering (KBQA): (1) for ranking generated queries in curated KBQA [201]; or (2) for measuring the quality of question paraphrases in open KBQA [25]. Our work is the first to adapt the idea to detect duplicate questions in PCQA websites.

To summarize, the main contributions of the chapter are:

- The first prototype of extracting knowledge from programming question answering communities: We design a process to extract knowledge from posts from Stack Overflow and form triples for the knowledge.
- Novel features for duplicate question detection: We represent questions as continuous vectors in a high dimensional space. We showed that neural embeddings capture semantic similarity better compared to traditional vector space representation such as tf-idf in the task of duplicate question detection.
- Association pairs for PCQA: We mined over 130K association pairs from known duplicate questions in Stack Overflow. The data contains frequently-occurring phrases that are domain-specific to PCQA. The source code for mining association pairs and mined pairs are publicly available for downloading.[4]
- Extensive experimental evaluation: In addition to Stack Overflow, we also tested PCQADup on other Stack Exchange sub-domains. Results suggest that PCQADup outperforms state-of-the-art benchmark by over than 30% in terms of recall for duplicate detection.

The rest of chapter is organized as follows. We describe a high level perspective of extracting knowledge from PCQA websites in Sect. 6.2. Section 6.3 details our proposed duplicate identification method, PCQADup. We report experimental results in Sect. 6.4. We review related work in Sect. 6.5 and finally discuss implications and caveats in Sect. 6.4.3.

6.2 Prototype of Knowledge Extraction from Programming Question Answering Communities

We propose a prototype of knowledge extraction from programming community based question answering websites, including the answer, tag and answer extraction.

6.2.1 Question Extraction

We extract the titles of the posts as the questions. We first identify duplicate posts, then extract question information using Part Of Speech (POS) tagging and dependency parsing. The details are as follows:

- Identify duplicate questions. We prune the duplicate questions using methods discussed in Sect. 6.3.
- Parse master questions. We parse the master questions and rewrite it to capture the meaning of the questions. Our method is inspired by Open IE work [25], where dependency parsing has been applied to detect query meaning. We discuss OPKE parsing process using an example question *"Java—How do I convert from int to String?"*: (1) remove language specific words if exist (e.g. *Java*), (2) parse dependencies of questions (using method in [202], see Fig. 6.1a), (3) identify subject (*I*), if subject is first person pronoun, then parse object part under the root in following steps (see the circled part in Fig. 6.1a), (4) identify root action (*covert*), (5) identify relationship ($\langle convert; fromint, tostring\rangle$) and (6) rewrite question (*"convert int to string"*). For more details, please refer to [25].

6.2.2 Answer and Tags Extraction

The accepted answer and tags can be obtained by considering two parameters '*AcceptedAnswerId*' and '*Tags*' respectively. We simply use the accepted answer as the final answer for a post. Although it might not be the best solution with highest votes, we believe that the questioner has the judgement on the solutions.

6.2.3 Triple Generation

After obtaining the questions and answers from previous steps, we generate triples with the format of $\langle$*"question"*; *"answer"*; *"tag1"*, *"tag2"*,...$\rangle$ where question and

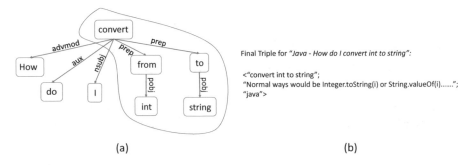

(a)   (b)

Fig. 6.1 Parse questions and generate triples. (**a**) Dependency parse tree. (**b**) Generated tuple

answers are obtained from Sect. 6.2.1 and tags are obtained from Sect. 6.2.2. So for the example question, the triple is depicted as in Fig. 6.1b.

6.3   Detecting Duplicate Posts in Programming QA Communities

The proposed duplicate identification method, PCQADup, has three components: (1) pre-processing; (2) feature modelling; and (3) classification.

6.3.1   Pre-processing

We pre-process text and clean the input as the first step. Text pre-processing is often necessary to standardise token representation. The pre-processing procedures involve both general text and programming language-specific processing.

- **Parsing and cleaning.** PCQA posts contain HTML tags, such as "<font>" and "<a>". We remove these HTML tags and retain only textual contents. We also remove posts with incomplete titles, as they are often not valid posts
- **Stop words pruning and tokenization.** We tokenize the sentences using Stanford Parser and prune stop words.[5]
- **Lemmatisation.** In natural language text, words exhibit morphological variations (e.g., singular vs. plural, present tense vs. past tense). We normalize the variations by lemmatization, e.g., converting "inlines" to "inline".
- **Symbol mapping.** For PCQA, posts often have symbols such as "$" (dollar) and "{" (brace) as they contain source code. From preliminary inspection we saw questions occasionally use the name of these symbols, e.g., *What is the purpose of the $ operator in a javascript function declaration?* vs. *Can someone explain the dollar sign in Javascript?*. As such, we map all symbols to their symbol names (i.e., "$" to "dollar"). We collect symbols and their respective names used in programming languages.[6] Note that we only map symbols that are tokenized as a single token, so e.g., "div[class=]" will not be converted. This conversion is performed only on question titles, as question bodies may contain source code where symbol conversion might not be sensible.

---

[5]Stanford Parser: http://nlp.stanford.edu/software/lex-parser.shtml; stop word list: http://snowball. tartarus.org/algorithms/english/stop.txt.

[6]Symbols are collected from: https://www.tutorialspoint.com/computer_programming/computer_ programming_characters.htm; the set of symbols used: {!, @, #, $, %, ^, &, *, (,), (),+, {, }, {}, >>, .*, _}.

6.3.2   Feature Modelling

We develop three types of features to detect duplicate questions. These features use both surface textual features (Association Score) and latent features (Vector Similarity and Topical Similarity) from the questions.

6.3.2.1   Vector Similarity

A conventional approach to vectorise text is to compute tf-idf scores for documents. We use tf-idf vectors to compute cosine similarity for a question pair, and the similarity serves as a baseline feature.

In the deep learning community, neural methods for learning word embeddings/vectors have seen plenty of successes for a range of NLP tasks. word2vec, a seminal work proposed by Mikolov et al. [180], is an efficient neural architecture to learn word embeddings via negative sampling using large corpora of text. Paragraph vectors, or doc2vec, was then introduced to extend word2vec to learn embeddings for word sequences [200]. doc2vec is agnostic to the granularity of the word sequence—it could learn embeddings for a sentence, paragraph or document. Two implementations of doc2vec were originally proposed: dbow and dmpv. The work in [203] evaluated dbow and dmpv on a range of extrinsic tasks and found that the simpler dbow trains faster and outperforms dmpv. We therefore use the dbow architecture for all experiments involving doc2vec in the paper. The hyperparameter settings used for training the doc2vec model is given in Table 6.1.

Given target question t (the new question being asked) and master question m (a previously answered question), the task is to classify whether t is a duplicate of m. We generate vectors for the pair's titles (t_m, t_t) and body contents (b_m, b_t). For the vectors of the concatenation of title and body, we do not concatenate the vectors of title and body. Instead, we generate vectors directly on the concatenation of title and body, denoted as c_m and c_t. This requires us to train the doc2vec models for title, body and title +body separately. We then compute cosine similarity for the pairs of vectors for (t_m, t_t), (b_m, b_t) and (c_m, c_t). As illustrated in Fig. 6.2, there are a total of five cosine similarity features for each question pair. Despite its simplicity, we find that these features are very effective for duplicate classification.

6.3.2.2   Topical Similarity

Latent Dirichlet Allocation (LDA) is a well-known implementation of topic model that extracts topics unsupervisedly from document collections. It posits that each document is a mixture of a small number of topics and each word's creation is attributable to one of the document's topic. However, LDA does not work well for short texts and as such may not be directly applicable to PCQA posts. To learn topics for our dataset, we adopt a modified LDA [204], which is designed for extracting topics from short texts.

Table 6.1 Hyper-parameter setting of doc2vec used

Parameters	Values	Description
Size	100	Dimension of word vectors
Window	15	Left/right context window size
min_count	1	Minimum frequency threshold for word types
Sample	1e-5	Threshold to down-sample high frequency words
Negative	5	Number of negative word samples
Iter	100	Number of training iterations

Fig. 6.2 Vector similarity features

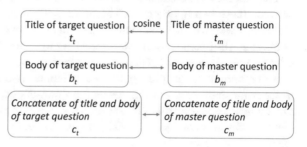

The output of the algorithm is two distributions, namely document-topic distribution and topic-words distribution. The document-topic distribution provides the probabilities of each document associated with each topic and the topic-words distribution presents the top words associated with each topic. To apply LDA, we consider each question as a document, and learn the distributions for all questions. We learn distributions separately over question titles, bodies and the concatenation of title and body for all questions, hence we obtain three document-topic distributions and three topic-words distributions. As the document-topic distribution contains numerical values (probabilities), these values are used to form vectors which represent the documents. Then we compute the cosine similarity between two vectors that represent two documents (i.e., questions) as topical similarity feature. For instance, all tiles are considered as a corpus, in which each title is a document. Then we apply LDA on this corpus and get document-topic distribution on titles.[7] We do the same operation for a corpus of bodies and then a corpus of title+body. Given a question pair, we compute the cosine similarities on vectors obtained from document-topic distributions over titles, bodies and title+body and obtain three values, which are topic similarity features.

6.3.2.3 Association Score

Manual inspection on duplicate classification using vector and topical similarity features reveals that these features fail to capture certain duplicates. For example,

---

[7]We set the number of words for a topic as 10 and set the number of topics as 30 for all experiments.

the pair *"Managing several EditTexts in Android"* and *"android: how to elegantly set many button IDs"* is not identified as duplicate. One possible reason could be that these features fail to identify that the phrases *"EditTexts"* and *"button IDs"* are related. This motivates us to look into mining association pairs—pair of phrases that co-occurs frequently in duplicate pairs—to further improve the performance of PCQADup.

We adopt the method developed by Berant and Liang [201] and Yin et al. [25] which uses association pairs to generate features for knowledge base question answering to improve recall rate. The intuition behind this method is that strong association between elements in target and master question suggests high similarity between these two questions. A key of strength of the method is its efficiency: it scales very well with large amount of data (in our case we consider tens of thousands of questions as potential duplicates for each target question). Note that we use only titles in a question pair to generate association pairs.

After mining association pairs, we generate association features and lexical features for another set of duplicates (and non-duplicates) and train a perception for duplicate classification. The aim of the exercise is to learn weights for these features. Given a new question pair, we use the trained perception to compute the association score for PCQADup.

Mining Association Pairs We first extract phrases that co-occur frequently in a duplicate pair in a PCQA corpus. To this end, we process over 25K duplicate questions on Stack Overflow. These questions are manually marked as duplicates by reputable users on Stack Overflow. Note that dataset used for mining association pairs is different to the dataset used for training the association score perception and PCQADup (dataset details in Sect. 6.4.1.1). For a pair of target question t and master question m, we learn alignment of words between t and m via machine translation [205]. The word alignment is performed in each direction of (m, t) and (t, m) and then combined using the grow-diag-final-and heuristic [206]. Given the word alignments, we then extract associated phrase pairs based on 3-gram heuristics developed by Och and Ney [207]. We prune pairs that occur less than 10 times in the data, and this produces over 130K association pairs. We henceforth refer to this set of association pairs as $\mathscr{A}$. Additionally, we include another complementary set of 1.3 million association pairs [201], denoted as $\mathscr{B}$, mined from 18 million pairs of question paraphrases from WikiAnswers.com. The rationale for including $\mathscr{B}$ is that it covers a wide range of general phrase pairs that could be relevant to our task. Table 6.2 presents examples of some association pairs.

Association Score Computation Given target question t and master question m, we iterate through all spans of text $s_t \in t$ and $s_m \in m$ and check if they are associated phrases in $\mathscr{A}$. If $\langle s_t, s_m \rangle \in \mathscr{A}$, we retrieve its counts from $\mathscr{A} \cup \mathscr{B}$ as the feature value, otherwise it is set to zero. We also generate lexical features for word pairs in t and m, checking for example if they share the same lemma, POS tag or are linked through a *derivation* link on WordNet [201, 208]. The full list of lexical and association features is presented in Table 6.3. Take Java related duplicate pairs for example, we generate over 80K association features in total using 5K randomly selected duplicate pairs based on Table 6.3. We then train a multilayer

Table 6.2 Examples of association pairs mined from stack overflow ($\mathscr{A}$) and WikiAnswers.com ($\mathscr{B}$)

Source	Phrase	Associated phrase
$\mathscr{A}$	Append	Concatenate
	Command prompt	Console
	Compiled executable	exe
	Java notify	Java use notifyall
$\mathscr{B}$	The primary language	Official language
	A triangular base	A triangle base

Table 6.3 Lexical and association features used in computing association score

Lexical features [25, 201]
$-lemma(s_m) \wedge lemma(s_t)$. $lemma(w)$ is the lemmatized word of w
$-pos(s_m) \wedge pos(s_t)$. $pos(w)$ is the POS tag of w
$-lemma(s_m)$ and $lemma(s_t)$ are synonyms?
$-lemma(s_m)$ and $lemma(s_t)$ are WordNet derivations?
Association pair counts as feature. $\mathscr{A} \cup \mathscr{B}$ is the set of mined association pairs
$-$count of $\langle lemma(s_m), lemma(s_t) \rangle$ if $\langle lemma(s_m), lemma(s_t) \rangle \in (\mathscr{A} \cup \mathscr{B})$, 0 otherwise

(m, t) is a pair of questions. s_m, s_t are spans in m and t respectively

perception with one hidden layer [209] using 5K duplicates and 5K non-duplicates for duplicate classification. The dataset used for generating association features is different to the dataset used for training the perception. The aim of the exercise is to learn weights for the 80K features. The weights learnt by the perception indicate the predictive power of the features. Features with zero weight are pruned from the feature space. This reduces the number of features to 16K. Example features are listed in Sect. 6.4.2.1. After obtaining weights for the features, we compute a weighted combination of the features for a given question pair (the pair is used to evaluate PCQADup) to generate a score:

$$score(m, t) = \sum_{i}^{N} v_{f_i} * \theta_{f_i}, \tag{6.1}$$

where N is the number of features with non-zero weights, v_{f_i} and θ_{f_i} are the value and weight of feature f_i respectively. This score constitutes the association score feature that feeds into PCQADup.

6.3.3 Binary Classification

We treat the duplication detection task as a binary classification problem. Given a pair of questions, the task is to classify whether they are duplicates or not. The two phases of classification in our work are described as follows.

Classifier Training Phase

To train a classifier (i.e., classification model), we first extract a set of features from the question pair. Three types of features are explored: (1) vector similarity; (2) topical similarity; and (3) association score. Feature generation is detailed in Sect. 6.3.2. In total, three features are generated for vector similarity, three features each for topical similarity and one feature for association score. In terms of classifiers, we experiment with the following models: decision tree [210], K-nearest neighbours (K-NN) [91], support vector machines (SVM) [211], logistic regression [212], random forest [213] and naive Bayes [214]. In terms of training data, the ratio of non-duplicate pairs (negative examples) to duplicate pairs (positive example) is very skewed. For example, in the Stack Overflow dump we processed (Sect. 6.4.1.1), 716,819 valid posts are tagged with "java" (case-insensitive), among which only 28,656 posts are marked as being duplicates. To create a more balanced dataset, we bias the distribution by under-sampling non-duplicate pairs [196, 197].

Duplicates Detection Phase

Given a question pair (m, t), where m is an existing question and t is a new posted question, the trained classifier predicts whether they are duplicate. To construct these question pairs, we conduct a naive filtering approach to filter out questions that belong to a different programming language or technique using tags. Tags are a mandatory input when posting a new question on Stack Overflow, and as such are a reliable indicator of the topic the question belongs to. Specifically, we prune existing questions that have no common tags with t, thus narrowing the search space for candidate duplicate questions considerably. We additionally filter out questions that have no answers. We then generate question pairs for t with all remainder questions, and compute features (will discuss in next section) for the classifier to predict the labels. The labels for (m, t) indicate whether t is a duplicate to m.

6.4 Experimental Evaluation and Discussions

In this section, we provide analyses to answer several questions: (1) what are the most impactful association features?; (2) what is the optimal combination of features for PCQADup?; (3) how does PCQADup perform comparing to the state-of-the-art benchmarks?; and (4) how robust is PCQADup—does it work for other PCQA domains?

6.4.1 Setup

6.4.1.1 Datasets

Our primary evaluation dataset is Stack Overflow. To test robustness of PCQADup, we additionally evaluate PCQADup on other sub-domains of Stack Exchange.

Stack Overflow The Stack Overflow dataset consists of 28,793,722 questions posted from April 2010 to June 2016.[8] We prune questions without answers ("AnswerCount"=0), producing 11,846,518 valid questions, among which, 250,710 pairs of questions are marked as duplicates (by examining the "LinkTypeId" parameter provided in the dataset). We use all these duplicate pairs to mine association pairs. These association pairs are used in computing association score feature for valid posts of all programming languages.

We focus our evaluation on Java related posts as it has the highest duplicate ratio in the dataset (Sect. 6.4.2.4) and obtain 28.6K duplicate question pairs with "Java" tag. As the ratio of non-duplicates to duplicates is very skewed, we keep all duplicates and randomly sample equal number of non-duplicates following the procedure described in Sect. 6.3.3. We thus obtain 28.6K non-duplicate question pairs. We use these pairs to generate features and training/test sets as follows: (1) For computing vector similarity feature, we use all these 114.4K = (28.6 + 28.6)*2 questions to train the doc2vec model. We do not train on all Java related posts due to the huge computation cost; (2) To generate topics for computing topical feature, we apply topic model on all these 114.4K questions; (3) In the computation of association score, we firstly use 5K duplicate pairs for generating features, then use different 5K duplicate pairs working with 5K non-duplicate pairs randomly selected from the 28.6K non-duplicate pairs to train the multilayer perception to learn weights of features; (4) Finally, we use question pairs that are not used in (3) for training and evaluating PCQADup. Specifically, we use 18.6K = 28.6-5-5 duplicate pairs and 18.6K non-duplicate pairs and split them into training and test set for the binary classification experiment in the ratio of 4:1.

The same splitting procedure is done to all other programming language datasets except for the ones with less than 10K duplicate questions because of insufficient data for evaluating PCQADup. Therefore, we split the duplicates of such datasets into the three parts of generating association features, learning association feature weights, and training/testing PCQADup classification performances using the ratio of 1:1:3. For the training/testing split, we follow the ratio of 4:1 as we did for Java related posts.

CQADupStack CQADupStack provides datasets for 12 sub-forums in Stack Exchange.[9] We select eight sub-forums that have texts of syntactic languages. Table 6.4 gives the number of total questions and the number of duplicate questions for the selected sub-forums. As the number of duplicates in these datasets is small, we generate only vector similarity and topical similarity features for the question pairs. We train the doc2vec model, which is used in the construction of vector similarity features, using all the posts. Then we use trained model to infer the vectors of the questions and compute cosine similarities of vectors. For classification experiments, we split ratio of 4:1 for training and test set as we do for Stack Overflow dataset.

---

[8] https://archive.org/details/stackexchange.

[9] http://nlp.cis.unimelb.edu.au/resources/cqadupstack/.

Table 6.4 CQADupStack
sub-forum statistics

Sub-forum	# of duplicates	# of all posts
Android	772	23,697
Gis	891	38,522
Mathematica	865	17,509
Programmers	1020	33,052
Stats	670	42,921
Unix	1113	48,454
Webmasters	529	17,911
Wordpress	549	49,146

6.4.1.2 Evaluation Metrics

We evaluate PCQADup using the following metrics:

- Recall and F_1 score. In this work, recall reflects the ability to identify duplicate pairs among the true duplicate pairs. F_1 is computed with both recall and precision and is a measure of accuracy.
- Area under the Receiver Operating Characteristic (ROC) curve (AUC). ROC curve shows how true positive rate changes with respect to false positive rate. AUC computes the probability that our model will rank a duplicate higher than a non-duplicate. Higher AUC score indicates better classification performance.

6.4.2 Results

6.4.2.1 Association Feature Weights Learner

To generate the association score for a question pair, we explore three learning algorithms that can asynchronously learn weights for a large number of features: (1) stochastic gradient descent optimized classifier [215]; (2) online passive aggressive [216]; and (3) perception with one hidden layer [209].

Table 6.5 presents performance of these algorithms in terms of classification accuracy. Perception achieves the best accuracy (bolded); we thus use the weights learned by perception to compute association score. Figure 6.3 illustrates the top 10 features weighted by perception. "l*" denotes left, "r*" denotes right. For example, "l_pos=DT N, r_pos=N" is a feature of a phrase pair, in which the POS tag of left/right phrase is "DT N"/"N". "Identity" means the phrases in the pair have the same lemma. "l_pos=a, r_pos=java" is the content feature of a phrase pair (a, java).

Table 6.5 Accuracy of different learning algorithms

Learning algorithm	F_1 score
Linear SVM with SGD	0.7825
On-line passive aggressive	0.7330
Perception with hidden variables	**0.8475**

Fig. 6.3 Relative importance of top 10 features

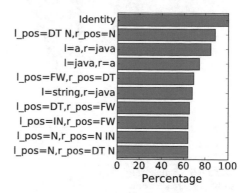

6.4.2.2 Feature Analyses

We analyse classification performance of PCQADup using following features: doc2vec vector similarity (VS), topical similarity (TP) and association score (ASS). We compare the performances of these features and their combinations with baseline feature VSM. We use random forest as the binary classifier in this experiment (comparisons of different classifiers are given later).

Figure 6.4 presents the ROC curve and AUC score when these features are used independently. We see that ASS performs the best, while VS is slightly worse and TP has the worst performance among the three. The baseline feature VSM using tf-idf scores generates lower AUC score than the three proposed features. The good performance of ASS indicates that the co-occurring phrase pairs between questions are important to reflect the duplicates. The relative worse performance of TP shows that topical information is difficult to learn in short PCQA posts. The baseline VSM performs only marginally worse than VS, revealing that conventional vectorisation method is not much worse than neural methods for our task.

Classification recall and F_1 score using different combinations of features are summarized in Table 6.6. As with the AUC scores, ASS is the most important feature when used independently. The combination of VS and TP performs worse than ASS alone, indicating the effectiveness of ASS feature. Combining vector similarity with association score, VS+ASS further improves performance of ASS by approximately 6% in recall and 4% in F_1 score. The combination of topical similarity and association score, TP+ASS, performs slightly better than VS+ASS. All the feature combinations performance better than the feature working independently in terms of recall, suggesting there is complementarity between these each two features.

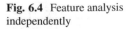

Fig. 6.4 Feature analysis
independently

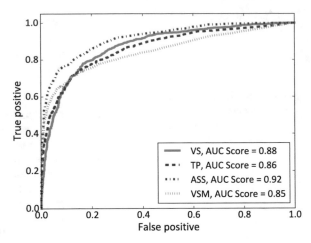

Table 6.6 Comparison
among different feature
combinations in PCQADup

Features	Recall	F_1 score
VSM	0.6570	0.7599
VS	0.7390	0.7845
TP	0.7430	0.7892
ASS	0.7760	0.8251
VS+TP	0.7450	0.8187
VS+ASS	0.8330	0.8695
TP+ASS	0.8490	0.8735
VS+TP+ASS	**0.8700**	**0.9067**

The bold means the best result

Therefore, the best performance is achieved by combining all features, namely
VS+TP+ASS yielding a further improvement of at least 2% to TP+ASS over all
metrics.

6.4.2.3   Classification Algorithms

We measure the classification performance using six classifiers: decision tree, K-
nearest neighbours, linear SVM, logistic regression, random forest and naive Bayes.
The basic parameters used in these classifiers are as follows: We set the maximum
depth of the tree to five and use Gini impurity to measure the tree split in decision
tree. We use $K = 5$ for K-nearest neighbours and weight each neighbourhood
equally. In SVM, we use linear kernel. For logistic regression, we use l_2 penalty. In
random forest, the number of trees is set to 10 and the maximum depth of the tree is
five. We use a Gaussian naive Bayes classifier in this work.

Figure 6.5 presents the ROC curves and AUC scores. From the figure, we can
see that our features perform very well with all six classifiers with little differences.
K-NN and naive Bayes are only marginally worse than decision tree, SVM and

Fig. 6.5 ROC curve on different classifiers on PCQADup (VS+TP+ASS)

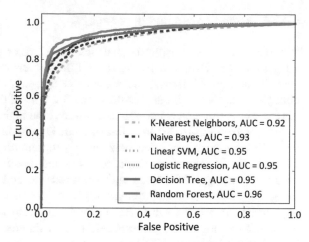

Table 6.7 Comparison among different classifiers on PCQADup (VS+TP+ASS)

Features	Recall	F_1 score
K-Nearest neighbours	0.7590	0.8272
Naive bayes	0.7700	0.8397
Linear SVM	0.8210	0.8725
Logistic regression	0.8340	0.8747
Decision tree	0.8280	0.8870
Random forest	**0.8720**	**0.9013**

The bold means the best result

logistic regression in terms of AUC scores. Random forest generates the highest AUC score. Table 6.7 presents recall and F_1 score performance for the classifiers. Here we see that random forest has the best performance in both recall and F_1 score. Logistic regression gives the second best recall, followed by decision tree, linear SVM and naive Bayes. K-nearest neighbours performs worst over the two metrics. The result that random forest is the best is consistent with the results obtained from extensive empirical evaluations from [217]. This is because random forest is an ensemble learning method that constructing a number of decision trees and uses the average value to improve the classification performance. SVM works well on high-dimensional data, but our feature vector is only nine-dimensional. This is a possible reason that SVM only gives mediocre performance. The probabilistic algorithms performs different in our evaluations: logistic regression gives good performance and is better than SVM, while the naive Bayes performs much worse. As naive Bayes assumes that the features are independent (not strictly), its undesirable performance indicating that the features (i.e., VS, TP, ASS) have strong dependencies on each other. Although K-NN is non-parametric, giving flexibility on the decision boundary, it assumes each feature has equal importance. The worst performance of K-NN in our evaluation shows that our features have different importance regard to the classification performance. We therefore use random forest as the classifier in all other evaluations.

6.4.2.4 Other Programming Languages

We first collect some duplicate statistics for 22 most popular programming languages on Stack Overflow. To aggregate different versions of a particular language, we collapse tags of different versions into one tag, e.g., "html" and "html5" tags are collapsed into "html". Figure 6.6 illustrates the ratio of duplicates to non-duplicates for these programming languages, and Java has the highest ratio of duplicates.

We compare PCQADup to state-of-the-art DupPredictor [198] and Dupee [197] on PCQA duplicate detection over seven programming languages. Performance values of DupPredictor and Dupee are obtained directly from [197]. The datasets used for comparing PCQADup with DupPredictor and Dupee are not exactly the same. For comparison, we form the duplicate pairs according to [197], hence the duplicate pairs are the same. But the non-duplicate pairs are randomly chosen independently. Figure 6.7 presents results over duplicate classification recall rate. For all seven languages, PCQADup outperforms DupPredictor and Dupee by at least 11% (Ruby). The biggest improvement is on C++, where the gain is over 35%. Although the numbers are not directly comparable due to the different non-duplicate part, the huge difference over recall rates are indicative that PCQADup has promising performance compared to DupPredictor and Dupee.

6.4.2.5 Performance on CQADupStack

To showcase the effectiveness of PCQADup on CQADupStack sub-forum datasets, we construct the TP features using only title-to-title vector similarity of posts and regard this feature as a baseline for comparison. Figure 6.8 shows the performance

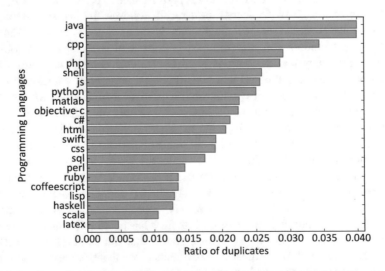

Fig. 6.6 Duplicate distribution of different programming languages (top-22 ranked by ratio)

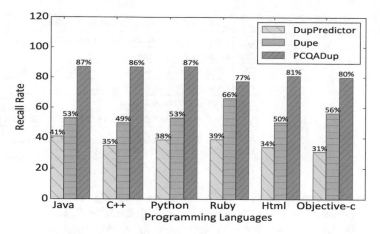

Fig. 6.7 Comparison to the state-of-the-art systems

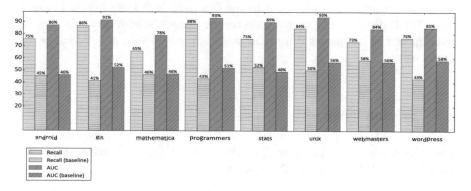

Fig. 6.8 Performance of PCQADup (VS+TP) on CQADupStack datasets

of PCQADup on eight selected sub-forums in CQADupStack over recall rate and AUC score. The recall rates of PCQADup are noticeably lower here compared to that of Stack Overflow. One reason is that we have much less duplicate and non-duplicate pairs trained in classification for CQADupStack. Another reason is that we use only VS and TP features for CQADupStack, as there is insufficient data to mine association pairs and generate association scores. Encouragingly, compared to the baseline feature, PCQADup is achieving at least 15% more recall rate (webmasters) and 27% more AUC score (wordpress) on most sub-forums.

6.4.3 Discussions

Despite the strong performance of PCQADup, there is room for improvement. We present some cases (and their reasons) where PCQADup does not successfully

Table 6.8 Examples of duplicate detection exceptions

Q_1	Causes of 'java.lang.NoSuchMethodError'
Q_2	How should a minimal Java program look like?
Reason: Implicit expression	
Q_3	What does (function (x,y)...)(a,b); mean in JavaScript?
Q_4	javascript syntax explanation
Reason: Difficult paraphrases	
Q_5	Finding whether the element exists in whole html page
Q_6	isset in Jquery?
Reason: Cross-language duplicates	

detect them as duplicates in Table 6.8. For example, Q_1 asks for solution for a specific problem "causes of java.lang.NoSuchMethodError". However, question issuer for Q_2 has problem related to "java.lang.NoSuchMethodError" but does not phrase explicitly in the question. Q_3 is phrased in a very specific manner, and has little word overlap with Q_4. Q_6 asks for functions in JQuery that is similar to "isset" in PHP which crosses the programming language boundary. We leave the development of techniques that tackle cross-language duplicate detection for our future work.

6.5 Related Work

Our work is related to previous studies in two fields: (1) question retrieval from QA communities; and (2) mining PCQA websites.

6.5.1 Question Retrieval from QA Communities

Large community question-answering data enabled studies to automatically retrieve response of existing questions to answer newly issued questions [191–196]. Cao et al. [193, 194] explored category information in Yahoo! Answers and combined a language model with a translation model to estimate the relevance of existing question-answer pairs to a query question. Shtok et al. [195] treated the question-answering task as a similarity identification problem, and retrieved questions that have similar titles as candidates. Wang et al. [191] identified similar questions by assessing the similarity of their syntactic parse trees. Yang et al. [196] classified questions based on heuristics and topical information to predict whether they will be answered. Zhou et al. [192] studied synonymy and polysemy to measure question similarity and built a concept thesaurus from Wikipedia for the task. A key difference of our work from these is that we are interested in identifying duplicates specifically for the PCQA domain.

6.5.2 Mining PCQA Websites

There has been plenty of research dedicated to mining information from PCQA to assist software development [197, 198, 218–220]. Ahasanuzzaman et al. [197] conducted an extensive analysis to understand why duplicate questions are created on Stack Overflow, and trained a binary classification model on question pairs to identify duplicate questions. Their work adopted features from [199] that mine duplicates from Twitter posts using textual and semantic features. Prior to [197], Zhang et al. [198] proposed to classify duplicate questions using title, description, topic and tag similarity as features. It was one of the first studies on duplicate detection in the PCQA domain. To better understand traits and behaviours of Stack Overflow users, Bazelli et al. [218] experimented with a linguistic inquiry and word count tool to analyse text written by users to predict their personalities and categorise them into several groups. Treude et al. [220] analysed the types of questions posted on Stack Overflow and examined which kind of questions are well answered and which ones are unanswered. Correa et al. [219] studied the characteristics of deleted questions on Stack Overflow to predict the deletion of a newly issued question. Our work also targets duplicate detection in PCQA websites, but develops novel features by leveraging association pairs and neural embeddings that largely outperforms the state-of-the-art works.

6.6 Summary

In this chapter, we have introduced PCQADup, a new state-of-the-art duplicate detection system for the PCQA domain. PCQADup is driven by few features, derived using methods from the deep learning and machine learning literature. When testing each feature independently, we found that vector representation learnt by neural models perform better than traditional tf-idf, although the difference is small. Combining all features in a classification model, PCQADup outperforms state-of-the-art duplicate detection systems by over 30% in multiple programming languages. As a product of the association score feature, we mined a set of associated phrases from duplicate questions on Stack Overflow. These phrases are domain-specific to PCQA, and could be used in other tasks such as keyword recommendation for forum searching.

Chapter 7
Conclusion

In this chapter, we summarize the contributions of this book and discuss some future research directions for data management of knowledge bases.

7.1 Summary

Knowledge base (KB) is one of the most essential components in realizing the idea of Semantic Web and has been recently receiving much research attention as a resource for providing knowledge, an auxiliary tool for facilitating the searching on search engines, an expert system for helping decision making in various domains. However, with the heterogeneous data sources and complex querying mechanism, this research area faces many challenges. Further research on effective and efficient data management approaches will be essential in establishing the future knowledge base management systems.

In this book, we address several research issues in data management in knowledge bases covering topics from knowledge retrieval and extraction. We propose approaches for optimizing and predicting the query performance from the querying interface of knowledge base. We examine two type of unstructured Web data sources for extraction knowledge. We also tackle the problem of data clustering. In particular, we summarize our main research contributions in the following:

- **Client-side cache based query optimization**. We proposed a SPARQL Endpoint Cache Framework, SECF [85, 86], to improve the overall querying performance on the SPARQL endpoints that are built upon knowledge bases. SECF utilizes machine learning techniques to learn clients' query patterns and suggests similar queries, whose results are prefetched and cached in order to reduce the overall querying time. We also developed a cache replacement algorithm, Modified Simple Exponential Smoothing, MSES [84], to replace the less valuable cache content, i.e., queries, with more valuable ones. MSES

© Springer International Publishing AG, part of Springer Nature 2018
W. E. Zhang, Q. Z. Sheng, *Managing Data From Knowledge Bases:
Querying and Extraction*, https://doi.org/10.1007/978-3-319-94935-2_7

outperforms the most used cache replacement algorithm LRU-2. Improved MSES further reduces the space overhead by considering a part of estimation records without losing cache hit rate.

- **Feature modelling via structural and syntactical information**. To leverage machine learning techniques in the query performance prediction, we proposed a template-based feature modelling using query's structural and syntactical information [85, 86, 112, 113]. This modelling approach eschew the need of knowing underlying databases and the query language. Moreover, it greatly outperforms the state-of-the-art method, cluster-based feature modelling method in terms of training and suggestion time. We then applied several widely adopted prediction models to predict the performance of SPARQL query. We considered the performance in both warm stage and cold stage of the system and the former one are never considered in previous works [112, 113]. We also performed extensive analysis of the real world queries from publicly accessible SPARQL endpoints and observed that many queries are issued by programmatic users, who tend to issue queries using query templates.

- **Data clustering via non-negative matrix factorization on stiefel manifold**. To perform more accurate and fast data clustering, we developed an algorithm called Nonlinear Riemannian Conjugate Gradient Orthogonal Non-negative Matrix Factorization, NRCG-ONMF [133]. NRCG-ONMF achieves higher clustering accuracy by imposing one of the factor matrices with orthogonality constraints and performing non-linear search on *Stiefel* manifold. The update of the two factor matrices is in an alternative way. We evaluated NRCG-ONMF on multiple real world datasets, the result of which demonstrate that the proposed NRCG-ONMF method outperforms other ONMF methods in terms of the effectiveness on preservation of orthogonality, optimization efficiency and clustering performance. This algorithm also sheds lights on an promising way to efficiently perform ONMF and shows great potential to handle large scale problems.

- **Source code topic extraction**. Unstructured data exist widely on the Web and to extract knowledge from which is a challenging issue for researchers. We focused on extracting knowledge from one type of unstructured data, the source code and started by extracting topics from Web repositories that holds source codes for a large number of projects. To tackle this issue, we developed Embeded Topic Extraction, EmbTE, based on word embedding techniques [169]. We also compared the proposed algorithm with LDA and NMF by measuring the topic coherence derived from the extracted topical terms. Moreover, we developed an automatic term selection algorithm that is able to identify the most contributory terms for the topic extraction from source code.

- **Duplicate detection from programming question-answering community**. Another type of unstructured data is the question posted to question-answering communities. We focused on the posts from programming question-answering communities, where questioners issued programming related questions. The first step to extract knowledge from these posts is to clean the posts, in which duplicate identification is of great importance as there exist a large amount of duplicate questions which would mislead the follow-up users. To solve this problem, we introduced Programming Community based Question Answering Duplicate Identification, PCQADup, a method that utilize both textual and latent features of the posts and achieve high identification accuracy [190, 221]. We evaluated PCQADup in multiple programming languages and many other forum posts who contain unstructured texts. PCQADup outperforms state-of-the-art duplicate detection systems by over 30%.

- **Large associated pairs mined from duplicate posts**. As a product of the association score feature, one of the three developed features used in PCQADup, we mine a large set of associated phrases from duplicate questions on Stack Overflow. These phrases are domain-specific to programming community based question-answering, and could be used in other tasks such as keyword recommendation for forum searching. To mine this set, we utilized the word-alignment of duplicate questions and applied a deep perception learning algorithm. Our approach is not limited to mining programming based posts, but is general and can be applied to identifying duplicates from other forum questions.

7.2   Future Directions

Although data management in knowledge bases attracted much attention recently, several research issues still need to be addressed. In particular, we identify the following directions for future research in data management in knowledge bases.

- **Complex knowledge retrieval**. In curated knowledge bases, complex queries aiming at retrieving knowledge with more constraints are supported by complex structured queries. While in open knowledge bases, where the knowledge is represented with less structured schema and without inference logic, the ability of answering complex queries induces great challenges. Although knowledge retrieval for complex queries is of significant importance, very limited effort has been put into tackling this research issue. Existing works rely on the solution of building new query-specific knowledge bases to handle complex queries. However, this is not a generic approach that is unlikely to be applied to other knowledge bases. As the open knowledge base is promising because of its flexibility, automatic creation and broader knowledge coverage, there exist high demand for more generic and efficient approaches to answer complex queries issued by knowledge consumers.
- **Natural language query on knowledge bases**. Knowledge base has been receiving much attention as a resource for providing knowledge and an expert system for helping decision making in various domains. The adoption of knowledge base is determined by the easiness of its usage, the accuracy of the knowledge it provided and the efficiency of the knowledge retrieval. Expert of querying languages could issue appropriate questions or keywords that match the knowledge in knowledge bases well and effectively obtain the answers. To make the knowledge base useful to a wider range of consumers who lack of specific

knowledge, approaches to retrieve knowledge without expertise are under great demands. Many research efforts focus on turning natural language questions or keywords to well-developed structured queries [50, 51]. Knowledge consumers hence could issue questions in a more natural way. However, challenges still exist due to the variation and evolvement of natural languages.

- **Knowledge extraction from unstructured data**. In the Big Data era, vast volume of data are generated, published and consumed everyday. Unstructured data has rapid growth in generation and adoption. Computer World states that unstructured data might account for more than 80% of all data in organizations [222]. Unstructured data are in the form of either text, audio or video. The challenges of extracting knowledge from unstructured data lie in its schema-less characteristic, which requires non-trivial effort to be understood by machines, leading to the difficulty of automatic extraction. Although we put one of the first attempts to extract knowledge from forum posts (this chapter), which is a kind of unstructured textual data, research attentions should be paid also on other kinds of unstructured data.

- **Open knowledge extraction for multilingual knowledge**. Curated knowledge bases such as DBpedia and YAGO provide multilingual knowledge including knowledge in both English and other languages. However, there is no complete work for automatically extracting non-English knowledge for open knowledge bases. Some preliminary attempts have been made (e.g., [223]), but no further works follow. Multilingual information not only provides complementary knowledge to English-dominant knowledge bases, but presents language specific knowledge, which attracts more adoption of knowledge bases from a wider range of consumers—not limited to the ones who can understand English. Therefore, the need for building more complete knowledge base calls for effective approaches to extract knowledge from multilingual knowledge source.

References

1. Michael Färber, Basil Ell, Carsten Menne, and Achim Rettinger. A Comparative Survey of DBpedia, Freebase, OpenCyc, Wikidata, and YAGO. *Semantic Web Journal*, 1(1):1–5, 2015.
2. Raf Guns. Tracing the Origins of the Semantic Web. *Journal of the American Society for Information Science and Technology*, 64(10):2173–2181, 2013.
3. Tim Berners-Lee, James Hendler, and Ora Lassila. The Semantic Web. *Scientific American*, 284(5):29–37, 2001.
4. https://www.techopedia.com/definition/2511/knowledge-base-klog.
5. Jens Lehmann, Robert Isele, Max Jakob, Anja Jentzsch, Dimitris Kontokostas, Pablo N. Mendes, Sebastian Hellmann, Mohamed Morsey, Patrick van Kleef, Sören Auer, and Christian Bizer. DBpedia - A Large-scale, Multilingual Knowledge Base Extracted from Wikipedia. *Semantic Web Journal*, 6(2):167–195, 2015.
6. Fabian M. Suchanek, Gjergji Kasneci, and Gerhard Weikum. Yago: A Core of Semantic Knowledge. In *Proc. of the 16th International World Wide Web Conference (WWW 2007)*, pages 697–706, Banff, Canada, May 2007.
7. https://www.wikidata.org/wiki/Wikidata:Main_Page.
8. Gene Ontology Consortium et al. The Gene Ontology Project in 2008. *Nucleic acids research*, 36(1):440–444, 2008.
9. George A Mille, Richard Beckwith, Christiane Fellbaum, Derek Gross, and Katherine J Miller. Introduction to WordNet: An Online Lexical Database. *International Journal of Lexicography*, 3(4):235–244, 1990.
10. https://twitter.com/jeffjarvis/status/783338071316135936.
11. http://blogs.bing.com/search/2013/03/21/understand-your-world-with-bing/.
12. https://www.fastcompany.com/3006389/where-are-they-now/entity-graph-facebook-calls-users-improve-its-search.
13. https://engineering.linkedin.com/blog/2016/10/building-the-linkedin-knowledge-graph.
14. Michael Martin, Jörg Unbehauen, and Sören Auer. Improving the Performance of Semantic Web Applications with SPARQL Query Caching. In *Proc. of the 7th Extended Semantic Web Conference (ESWC 2010)*, pages 304–318, Heraklion, Crete, Greece, 2010.
15. Yanfeng Shu, Michael Compton, Heiko Müller, and Kerry Taylor. Towards Content-Aware SPARQL Query Caching for Semantic Web Applications. In *Proc. of the 14th International Conference on Web Information Systems Engineering (WISE 2013)*, pages 320–329, Nanjing, China, October 2013.
16. Mengdong Yang and Gang Wu. Caching Intermediate Result of SPARQL Queries. In *Proc. of the 20th International World Wide Web Conference (WWW 2011)*, pages 159–160, Hyderabad, India, March 2011.

© Springer International Publishing AG, part of Springer Nature 2018   127
W. E. Zhang, Q. Z. Sheng, *Managing Data From Knowledge Bases:*
Querying and Extraction, https://doi.org/10.1007/978-3-319-94935-2

17. Nikolaos Papailiou, Dimitrios Tsoumakos, Panagiotis Karras, and Nectarios Koziris. Graph-Aware, Workload-Adaptive SPARQL Query Caching. In *Proc. of the International Conference on Management of Data (SIGMOD 2015)*, pages 1777–1792, Melbourne, Australia, May 2015.
18. Ralph Grishman. *Information Extraction: Techniques and Challenges*. Springer, 1997.
19. Smith Reid. Knowledge-Based Systems Concepts, Techniques, Examples. http://www.reidgsmith.com/, 1985.
20. UniProt Consortium et al. UniProt: A Hub for Protein Information. *Nucleic acids research*, page gku989, 2014.
21. http://lod.geospecies.org/.
22. https://agclass.nal.usda.gov/.
23. Kurt D. Bollacker, Colin Evans, Praveen Paritosh, Tim Sturge, and Jamie Taylor. Freebase: A Collaboratively Created Graph Database for Structuring Human Knowledge. In *Proc. of the ACM SIGMOD International Conference on Management of Data (SIGMOD 2008)*, pages 1247–1250, Vancouver, Canada, June 2008.
24. Douglas B. Lenat. CYC: A Large-Scale Investment in Knowledge Infrastructure. *Communnications of ACM*, 38(11):32–38, 1995.
25. Pengcheng Yin, Nan Duan, Ben Kao, Jun-Wei Bao, and Ming Zhou. Answering Questions with Complex Semantic Constraints on Open Knowledge Bases. In *Proc. of the 24th ACM International Conference on Information and Knowledge Management (CIKM 2015)*, pages 1301–1310, Melbourne, VIC, Australia, October 2015.
26. Oren Etzioni, Michael J. Cafarella, Doug Downey, Stanley Kok, Ana-Maria Popescu, Tal Shaked, Stephen Soderland, Daniel S. Weld, and Alexander Yates. Web-scale Information Extraction in KnowItAll. In *Proc. of the 13th international conference on World Wide Web (WWW 2014)*, pages 100–110, New York, NY, USA, May 2004.
27. Anthony Fader, Luke S. Zettlemoyer, and Oren Etzioni. Paraphrase-Driven Learning for Open Question Answering. In *Proc. of the 51st Annual Meeting of the Association for Computational Linguistics (ACL 2013)*, pages 1608–1618, Sofia, Bulgaria, August 2013.
28. Anthony Fader, Luke Zettlemoyer, and Oren Etzioni. Open Question Answering over Curated and Extracted Knowledge Bases. In *Proc. of the 20th ACM SIGKDD International Conference on Knowledge Discovery and Data Mining (KDD 2014)*, pages 1156–1165, New York, NY, USA, August 2014.
29. David A. Ferrucci, Eric W. Brown, Jennifer Chu-Carroll, James Fan, David Gondek, Aditya Kalyanpur, Adam Lally, J. William Murdock, Eric Nyberg, John M. Prager, Nico Schlaefer, and Christopher A. Welty. Building Watson: An Overview of the DeepQA Project. *AI Magazine*, 31(3):59–79, 2010.
30. Mahnoosh Kholghi, Laurianne Sitbon, Guido Zuccon, and Anthony N. Nguyen. External Knowledge and Query Strategies in Active Learning: a Study in Clinical Information Extraction. In *Proc. of the 24th ACM International Conference on Information and Knowledge Management (CIKM 2015)*, pages 143–152, Melbourne, VIC, Australia, October.
31. Mark Dredze, Paul McNamee, Delip Rao, Adam Gerber, and Tim Finin. Entity disambiguation for knowledge base population. In *Proc. of the 23rd International Conference on Computational Linguistics (COLING 2010)*, pages 277–285, Osaka, Japan, December 2010.
32. Tatiana N. Erekhinskaya, Meghana Satpute, and Dan I. Moldovan. Multilingual eXtended WordNet Knowledge Base: Semantic Parsing and Translation of Glosses. In *Proc. of the Ninth International Conference on Language Resources and Evaluation, (LREC 2014)*, pages 2990–2994, Reykjavik, Iceland, May 2014.
33. Michael Cochez, Stefan Decker, and Eric Prud'hommeaux. Knowledge Representation on the Web Revisited: The Case for Prototypes. In *Proc. of the 15th International Semantic Web Conference (ISWC 2016)*, pages 151–166, Kobe, Japan, October 2016.
34. Wei Shen, Jianyong Wang, and Jiawei Han. Entity Linking with a Knowledge Base: Issues, Techniques, and Solutions. *IEEE Transactions on Knowledge and Data Engineering*, 27(2):443–460, 2015.

35. Zhuoyu Wei, Jun Zhao, Kang Liu, Zhenyu Qi, Zhengya Sun, and Guanhua Tian. Large-scale Knowledge Base Completion: Inferring via Grounding Network Sampling over Selected Instances. In *Proc. of the 24th ACM International Conference on Information and Knowledge Management (CIKM 2015)*, pages 1331–1340, Melbourne, VIC, Australia, October 2015.

36. Huan Sun, Hao Ma, Wen-tau Yih, Chen-Tse Tsai, Jingjing Liu, and Ming-Wei Chang. Open Domain Question Answering via Semantic Enrichment. In *Proc. of the 24th International Conference on World Wide Web (WWW 2015)*, pages 1045–1055, Florence, Italy, May 2015.

37. Farzaneh Mahdisoltani, Joanna Biega, and Fabian M. Suchanek. YAGO3: A Knowledge Base from Multilingual Wikipedias. In *Proc. of the 17th Biennial Conference on Innovative Data Systems Research (CIDR 2015)*, Asilomar, CA, USA, January 2015.

38. Olena Medelyan, David N. Milne, Catherine Legg, and Ian H. Witten. Mining Meaning from Wikipedia. *International Journal of Human-Computer Studies*, 67(9):716–754, 2009.

39. Eytan Adar, Michael Skinner, and Daniel S. Weld. Information Arbitrage Across Multilingual Wikipedia. In *Proc. of the 2nd International Conference on Web Search and Web Data Mining (WSDM 2009)*, pages 94–103, Barcelona, Spain, February 2009.

40. Thanh Hoang Nguyen, Viviane Pereira Moreira, Huong Nguyen, Hoa Nguyen, and Juliana Freire. Multilingual Schema Matching for Wikipedia Infoboxes. *PVLDB*, 5(2):133–144, 2011.

41. Daniel Rinser, Dustin Lange, and Felix Naumann. Cross-lingual Entity Matching and Infobox Alignment in Wikipedia. *Information Systems*, 38(6):887–907, 2013.

42. Zhichun Wang, Juanzi Li, Zhigang Wang, and Jie Tang. Cross-lingual Knowledge Linking Across Wiki Knowledge Bases. In *Proc. of the 21st World Wide Web Conference (WWW 2012)*, pages 459–468, Lyon, France, April 2012.

43. Oren Etzioni, Anthony Fader, Janara Christensen, Stephen Soderland, and Mausam. Open information extraction: The second generation. In *Proc. of the 22nd International Joint Conference on Artificial Intelligence (IJCAI 2011)*, pages 3–10, Barcelona, Spain, July 2011.

44. Michele Banko, Michael J. Cafarella, Stephen Soderland, Matthew Broadhead, and Oren Etzioni. Open Information Extraction from the Web. In *Proc. of the 20th International Joint Conference on Artificial Intelligence (IJCAI 2007)*, pages 2670–2676, Hyderabad, India, January 2007.

45. Jun Zhu, Zaiqing Nie, Xiaojiang Liu, Bo Zhang, and Ji-Rong Wen. StatSnowball: A Statistical Approach to Extracting Entity Relationships. In *Proc. of the 18th International Conference on World Wide Web (WWW 2009)*, pages 101–110, Madrid, Spain, April 2009.

46. Fei Wu and Daniel S. Weld. Open Information Extraction Using Wikipedia. In *Proc. of the 48th Annual Meeting of the Association for Computational Linguistics (ACL 2010)*, pages 118–127, Uppsala, Sweden, July 2010.

47. Mausam, Michael Schmitz, Stephen Soderland, Robert Bart, and Oren Etzioni. Open Language Learning for Information Extraction. In *Proc. of the 2012 Joint Conference on Empirical Methods in Natural Language Processing and Computational Natural Language Learning (EMNLP-CoNLL 2012)*, pages 523–534, Jeju Island, Korea, July 2012.

48. Ana-Maria Popescu, Oren Etzioni, and Henry A. Kautz. Towards a Theory of Natural Language Interfaces to Databases. In *Proc. of the 2003 International Conference on Intelligent User Interfaces (IUI 2003)*, pages 149–157, Miami, FL, USA, January 2003.

49. Yuangui Lei, Victoria S. Uren, and Enrico Motta. SemSearch: A Search Engine for the Semantic Web. In *Proc. of the 15th International Conference on Managing Knowledge in a World of Networks (EKAW 2006)*, pages 238–245, Podebrady, Czech Republic, October 2006.

50. Qi Zhou, Chong Wang, Miao Xiong, Haofen Wang, and Yong Yu. SPARK: Adapting Keyword Query to Semantic Search. In *Proc. of the 6th International Semantic Web Conference (ISWC 2007)*, pages 694–707, Busan, Korea, November 2007.

51. Jens Lehmann and Lorenz Bühmann. AutoSPARQL: Let Users Query Your Knowledge Base. In *Proc. of the 8th Extended Semantic Web Conference (ESWC 2011)*, pages 63–79, Heraklion, Crete, Greece, May 2011.

52. Christina Unger, Lorenz Bühmann, Jens Lehmann, Axel-Cyrille Ngonga Ngomo, Daniel Gerber, and Philipp Cimiano. Template-based Question Answering Over RDF Data. In *Proc. of the 21st International World Wide Web Conference (WWW 2012)*, pages 639–648, Lyon, France, April 2012.

53. Mohamed Yahya, Klaus Berberich, Shady Elbassuoni, Maya Ramanath, Volker Tresp, and Gerhard Weikum. Natural Language Questions for the Web of Data. In *Proc. of the 2012 Joint Conference on Empirical Methods in Natural Language Processing and Computational Natural Language Learning (EMNLP-CoNLL 2012)*, pages 379–390, Jeju Island, Korea, July 2012.

54. Lei Zou, Ruizhe Huang, Haixun Wang, Jeffrey Xu Yu, Wenqiang He, and Dongyan Zhao. Natural Language Question Answering over RDF: A Graph Data Driven Approach. In *Proc. of the International Conference on Management of Data (SIGMOD 2014)*, pages 313–324, Snowbird, UT, USA, June 2014.

55. Camille Pradel, Ollivier Haemmerlé, and Nathalie Hernandez. Natural Language Query Interpretation into SPARQL using Patterns. In *Proc. of the Fourth International Conference on Consuming Linked Data (COLD 2013)*, pages 13–24, Sydney, Australia, October 2013.

56. Malte Sander, Ulli Waltinger, Mikhail Roshchin, and Thomas Runkler. Ontology-based Translation of Natural Language Queries to SPARQL. In *NLABD: AAAI Fall Symposium*, 2014.

57. Christina Unger and Philipp Cimiano. Pythia: Compositional Meaning Construction for Ontology-Based Question Answering on the Semantic Web. In *Proc. of the 16th International Conference on Applications of Natural Language to Information Systems (NLDB 2011)*, pages 153–160, Alicante, Spain, June 2011.

58. Danica Damljanovic, Milan Agatonovic, and Hamish Cunningham. FREyA: An Interactive Way of Querying Linked Data Using Natural Language. In *The Semantic Web: ESWC 2011 Workshops*, pages 125–138, Heraklion, Greece, May 2011.

59. http://www.w3.org/TR/rdf-sparql-query/.

60. http://en.wikipedia.org/wiki/SPARQL/.

61. https://www.w3.org/TR/rdf-sparql-protocol/.

62. Jeen Broekstra, Arjohn Kampman, and Frank van Harmelen. Sesame: A Generic Architecture for Storing and Querying RDF and RDF Schema. In *Proc. of the 1st International Semantic Web Conference (ISWC 2002)*, pages 54–68, Sardinia, Italy, June 2002.

63. Thomas Neumann and Gerhard Weikum. The RDF-3X Engine for Scalable Management of RDF Data. *The VLDB Journal*, 19(1):91–113, 2010.

64. Kevin Wilkinson. Jena Property Table Implementation. In *Proc. of the 2nd Scalable Semantic Web Knowledge Base Systems workshop in ISWC (SSWS 2006)*, Athens, GA, USA, November 2006.

65. Medha Atre, Jagannathan Srinivasan, and James A. Hendler. Bitmat: A main-memory bit matrix of rdf triples for conjunctive triple pattern queries. In *International Semantic Web Conference (Posters & Demos)*, 2008.

66. Medha Atre, Vineet Chaoji, Mohammed J. Zaki, and James A. Hendler. Matrix "Bit" Loaded: A Scalable Lightweight Join Query Processor for RDF Data. In *Proc. of the 19th International World Wide Web Conference (WWW 2010)*, pages 41–50, Raleigh, North Carolina, USA, April 2010.

67. Michael Stonebraker, Daniel J. Abadi, Adam Batkin, Xuedong Chen, Mitch Cherniack, Miguel Ferreira, Edmond Lau, Amerson Lin, Samuel Madden, Elizabeth J. O'Neil, Patrick E. O'Neil, Alex Rasin, Nga Tran, and Stanley B. Zdonik. C-Store: A Column-oriented DBMS. In *Proc. of the 31st International Conference on Very Large Data Bases (VLDB 2005)*, pages 553–564, Trondheim, Norway, September 2005.

68. Daniel J. Abadi, Adam Marcus, Samuel Madden, and Kate Hollenbach. SW-Store: A Vertically Partitioned DBMS for Semantic Web Data Management. *The VLDB Journal*, 18(2):385–406, 2009.

69. Daniel J. Abadi, Adam Marcus, Samuel Madden, and Katherine J. Hollenbach. Scalable Semantic Web Data Management Using Vertical Partitioning. In *Proc. of the 33rd International Conference on Very Large Databases (VLDB 2007)*, pages 411–422, Vienna, Austria, September 2007.
70. Valerie Bönström, Annika Hinze, and Heinz Schweppe. Storing RDF as a Graph. In *Proc. of the 1st Latin American Web Congress (LA-WEB 2003)*, pages 27–36, Sanitago, Chile, November 2003.
71. Lei Zou, Jinghui Mo, Lei Chen, M. Tamer Özsu, and Dongyan Zhao. gStore: Answering SPARQL Queries via Subgraph Matching. *The VLDB Endowment (PVLDB)*, 4(8):482–493, 2011.
72. Kai Zeng, Jiacheng Yang, Haixun Wang, Bin Shao, and Zhongyuan Wang. A Distributed Graph Engine for Web Scale RDF Data. *The VLDB Endowment (PVLDB)*, 6(4):265–276, 2013.
73. Kurt Rohloff and Richard E. Schantz. High-performance, massively scalable distributed systems using the mapreduce software framework: the SHARD triple-store. In *Proc. of SPLASH Workshop on Programming Support Innovations for Emerging Distributed Applications (PSI EtA)*, page 4, Reno, Nevada, USA, October 2010.
74. Orri Erling and Ivan Mikhailov. Virtuoso: RDF Support in a Native RDBMS. In *Semantic Web Information Management*, pages 501–519. 2009.
75. Li Ma, Zhong Su, Yue Pan, Li Zhang, and Tao Liu. RStar: an RDF storage and query system for enterprise resource management. In *Proc. of the 2004 ACM International Conference on Information and Knowledge Management (CIKM 2004)*, pages 484–491, Washington, DC, USA, November 2004.
76. Roberto De Virgilio, Fausto Giunchiglia, and Letizia Tanca, editors. *Semantic Web Information Management - A Model-Based Perspective*. Springer, 2009.
77. Octavian Udrea, Andrea Pugliese, and V. S. Subrahmanian. GRIN: A graph based RDF index. In *Proc. of the 22nd AAAI Conference on Artificial Intelligence (AAAI 2007)*, pages 1465–1470, Vancouver, British Columbia, Canada, July 2007.
78. Orri Erling and Ivan Mikhailov. Towards web scale RDF. 2008.
79. Andreas Harth and Stefan Decker. Optimized Index Structures for Querying RDF from the Web. In *Proc. of the 3rd Latin American Web Congress (LA-Web 2005)*, pages 71–80, Buenos Aires, Argentina, November 2005.
80. Pingpeng Yuan, Pu Liu, Buwen Wu, Hai Jin, Wenya Zhang, and Ling Liu. TripleBit: a Fast and Compact System for Large Scale RDF Data. *The VLDB Endowment (PVLDB)*, 6(7):517–528, 2013.
81. Olaf Hartig and Frank Huber. A Main Memory Index Structure to Query Linked Data. In *Proc. of the Workshop on Linked Data on the Web (LDOW 2011)*, Hyderabad, India, March 2011.
82. David Wood, Paul Gearon, and Tom Adams. Kowari: A platform for semantic web storage and analysis. In *XTech 2005 Conference*, pages 05–0402, 2005.
83. Andreas Harth, Jürgen Umbrich, Aidan Hogan, and Stefan Decker. YARS2: A Federated Repository for Querying Graph Structured Data from the Web. In *Proc. of the 6th International Semantic Web Conference (ISWC 2007), 2nd Asian Semantic Web Conference (ASWC 2007)*, pages 211–224, Busan, Korea, November 2007.
84. Wei Emma Zhang, Quan Z. Sheng, Kerry Taylor, and Yongrui Qin. Identifying and Caching Hot Triples for Efficient RDF Query Processing. In *Proc. of the 20th International Conference on Database Systems for Advanced Applications (DASFAA 2015)*, pages 259–274, Hanoi, Vietnam, April 2015.
85. Wei Emma Zhang, Quan Z. Sheng, Yongrui Qin, Lina Yao, Ali Shemshadi, and Kerry L. Taylor. SECF:Improving SPARQL Querying Performance with Proactive Fetching and Caching. In *Proc. of the 31st Annual ACM Symposium on Applied Computing (SAC 2016)*, pages 362–367, Pisa, Italy, April 2016.
86. Wei Emma Zhang, Quan Z. Sheng, Lina Yao, Kerry Taylor, Ali Shemshadi, and Yongrui Qin. A Learning Based Framework for Improving Querying on Web Interfaces of Curated Knowledge Bases. *ACM Transactions on Internet Technology (TOIT)*, 18(6):35, 2018.

87. Johannes Lorey and Felix Naumann. Detecting SPARQL Query Templates for Data Prefetching. In *Proc. of the 10th Extended Semantic Web Conference (ESWC 2013)*, pages 124–139, Montpellier, France, May 2013.

88. Ruben Verborgh, Olaf Hartig, Ben De Meester, Gerald Haesendonck, Laurens De Vocht, Miel Vander Sande, Richard Cyganiak, Pieter Colpaert, Erik Mannens, and Rik Van de Walle. Querying datasets on the web with high availability. In *Proc. of the 13th International Semantic Web Conference (ISWC 2014)*, pages 180–196, Riva del Garda, Italy, October 2014.

89. Qun Ren, Margaret H. Dunham, and Vijay Kumar. Semantic caching and query processing. *IEEE Transactions on Knowledge and Data Engineering*, 15(1):192–210, 2003.

90. Rakebul Hasan. Predicting SPARQL Query Performance and Explaining Linked Data. In *Proc. of the 11th Extended Semantic Web Conference (ESWC 2014)*, pages 795–805, Anissaras, Crete, Greece, May 2014.

91. Naomi S Altman. An Introduction to Kernel and Nearest-Neighbor Nonparametric Regression. *The American Statistician*, 46(3):175–185, 1992.

92. Peter J. Denning. The Working Set Model for Program Behaviour. *Communications of the ACM*, 11(5):323–333, 1968.

93. Elizabeth J. O'Neil, Patrick E. O'Neil, and Gerhard Weikum. The LRU-K Page Replacement Algorithm For Database Disk Buffering. In *Proc. of the International Conference on Management of Data (SIGMOD 1993)*, pages 297–306, Washington, D.C., USA, May 1993.

94. Nimrod Megiddo and Dharmendra S. Modha. ARC: A Self-Tuning, Low Overhead Replacement Cache. In *Proc. of the Conference on File and Storage Technologies (FAST 2003)*, San Francisco, California, USA, March 2003.

95. Justin J. Levandoski, Per-Åke Larson, and Radu Stoica. Identifying Hot and Cold Data in Main-Memory Databases. In *Proc. of 29th International Conference on Data Engineering (ICDE 2013)*, pages 26–37, Brisbane, Australia, April 2013.

96. Jorge Pérez, Marcelo Arenas, and Claudio Gutierrez. Semantics and Complexity of SPARQL. *ACM Transactions on Database Systems*, 34(3), 2009.

97. Alberto Sanfeliu and King-Sun Fu. A Distance Measure between Attributed Relational Graphs for Pattern Recognition. *IEEE Transactions on Systems, Man, and Cybernetics*, 13(3):353–362, 1983.

98. Leonard Kaufman and Peter Rousseeuw. *Clustering by Means of Medoids*. North-Holland, 1987.

99. Mohamed Morsey, Jens Lehmann, Sören Auer, and Axel-Cyrille Ngonga Ngomo. Usage-Centric Benchmarking of RDF Triple Stores. In *Proc. of the 26th AAAI Conference on Artificial Intelligence (AAAI 2012).*, Toronto, Canada, July 2012.

100. Harold Hotelling. Relations between two sets of variates. *Biometrika*, pages 321–377, 1936.

101. Ian Jolliffe. *Principal Component Analysis*. Wiley Online Library, 2002.

102. Daniel D Lee and H Sebastian Seung. Learning the parts of objects by non-negative matrix factorization. *Nature*, 401(6755):788–791, 1999.

103. Everette S Gardner. Exponential Smoothing: The State of The Art–Part II. *International Journal of Forecasting*, 22(4):637–666, 2006.

104. Javier R Movellan. A Quickie on Exponential Smoothing. 2007.

105. Norman L. Johnson, Adrienne W. Kemp, and Samuel Kotz. *Univariate Discrete Distributions (2nd Edition)*. Wiley, 1993.

106. Mengmeng Liu, Zachary G. Ives, and Boon Thau Loo. Enabling Incremental Query Re-Optimization. In *Proc. of the 2016 International Conference on Management of Data (SIGMOD 2016)*, pages 1705–1720, San Francisco, CA, USA, June 2016.

107. Parke Godfrey and Jarek Gryz. Answering Queries by Semantic Caches. In *Proc. of the 10th International Conference on Database and Expert Systems Applications (DEXA 1999)*, pages 485–498, Florence, Italy, August 1999.

108. Shaul Dar, Michael J. Franklin, Björn THór Jónsson, Divesh Srivastava, and Michael Tan. Semantic Data Caching and Replacement. In *Proc. of the 22rd International Conference on Very Large Data Bases (VLDB1996)*, pages 330–341, Mumbai (Bombay), India, September 1996.

109. Huanhuan Cao, Daxin Jiang, Jian Pei, Qi He, Zhen Liao, Enhong Chen, and Hang Li. Context-aware Query Suggestion by Mining Click-through and Session Data. In *Proc. of the 14th ACM SIGKDD Conference on Knowledge Discovery and Data Mining (KDD 2008)*, pages 875–883, Las Vegas, Nevada, USA, August 2008.

110. Géraud Fokou, Stéphane Jean, Allel Hadjali, and Mickaël Baron. Cooperative Techniques for SPARQL Query Relaxation in RDF Databases. In *Proc. of the 12th Extended Semantic Web Conference (ESWC 2015)*, pages 237–252, Portoroz, Slovenia, June 2015.

111. Shady Elbassuoni, Maya Ramanath, and Gerhard Weikum. Query Relaxation for Entity-Relationship Search. In *Proc. of the 8th Extended Semantic Web Conference (ESWC 2011)*, pages 62–76, Heraklion, Crete, Greece, May 2011.

112. Wei Emma Zhang, Quan Z. Sheng, Kerry L. Taylor, Yongrui Qin, and Lina Yao. Learning-Based SPARQL Query Performance Prediction. In *Proc. of the 17th International Conference on Web Information Systems Engineering (WISE 2016)*, pages 313–327, Shanghai, China, November 2016.

113. Wei Emma Zhang, Quan Z. Sheng, Yongrui Qin, Kerry Taylor, and Lina Yao. Learning-based SPARQL Query Performance Modeling and Prediction. *World Wide Web Journal*, 21(4): 1015–1035, 2018.

114. Mumtaz Ahmad, Songyun Duan, Ashraf Aboulnaga, and Shivnath Babu. Predicting completion times of batch query workloads using interaction-aware models and simulation. In *Proc. of the 14th International Conference on Extending Database Technology (EDBT 2011)*, pages 449–460, Uppsala, Sweden, March 2011.

115. Jennie Duggan, Ugur Çetintemel, Olga Papaemmanouil, and Eli Upfal. Performance prediction for concurrent database workloads. In *Proc. of the 2011 International Conference on Management of Data (SIGMOD 2011)*, pages 337–348, Athens, Greece, May 2011.

116. Jiexing Li, Arnd Christian König, Vivek R. Narasayya, and Surajit Chaudhuri. Robust Estimation of Resource Consumption for SQL Queries using Statistical Techniques. *The VLDB Endowment (PVLDB)*, 5(11):1555–1566, 2012.

117. Wentao Wu, Yun Chi, Shenghuo Zhu, Jun'ichi Tatemura, Hakan Hacigümüs, and Jeffrey F. Naughton. Predicting query execution time: Are optimizer cost models really unusable? In *Proc. of the 29th International Conference on Data Engineering (ICDE 2013)*, pages 1081–1092, Brisbane Australia, April 2013.

118. Mert Akdere, Ugur Çetintemel, Matteo Riondato, Eli Upfal, and Stanley B. Zdonik. Learning-based query performance modeling and prediction. In *Proc. of the 28th International Conference on Data Engineering (ICDE 2012)*, pages 390–401, Washington DC, USA, April 2012.

119. Archana Ganapathi, Harumi A. Kuno, Umeshwar Dayal, Janet L. Wiener, Armando Fox, Michael I. Jordan, and David A. Patterson. Predicting Multiple Metrics for Queries: Better Decisions Enabled by Machine Learning. In *Proc. of the 25th International Conference on Data Engineering (ICDE 2009)*, pages 592–603, Shanghai China, March 2009.

120. Sean Tozer, Tim Brecht, and Ashraf Aboulnaga. Q-Cop: Avoiding bad query mixes to minimize client timeouts under heavy loads. In *Proc. of the 26th International Conference on Data Engineering (ICDE 2010)*, pages 397–408, Long Beach, USA, March 2010.

121. Petros Tsialiamanis, Lefteris Sidirourgos, Irini Fundulaki, Vassilis Christophides, and Peter A. Boncz. Heuristics-based query optimisation for SPARQL. In *Proc. of the 15th International Conference on Extending Database Technology (EDBT 2012)*, pages 324–335, Uppsala, Sweden, March 2012.

122. Markus Stocker, Andy Seaborne, Abraham Bernstein, Christoph Kiefer, and Dave Reynolds. SPARQL Basic Graph Pattern Optimization Using Selectivity Estimation. In *Proc. of the 17th International World Wide Web Conference (WWW 2008)*, pages 595–604, Beijing, China, April 2008.

123. Anand Rajaraman and Jeffrey David Ullman. *Mining of Massive Datasets*. Cambridge University Press, 2011.

124. Alex Smola and Vladimir Vapnik. Support Vector Regression Machines. *Advances in neural information processing systems*, 9:155–161, 1997.

125. Gareth James, Daniela Witten, Trevor Hastie, and Robert Tibshirani. *An Introduction to Statistical Learning*. Springer, 2013.
126. Damian Bursztyn, François Goasdoué, and Ioana Manolescu. Optimizing reformulation-based query answering in RDF. In *Proc. of the 18th International Conference on Extending Database Technology (EDBT 2015)*, pages 265–276, Brussels, Belgium, March 2015.
127. Xindong Wu, Vipin Kumar, J. Ross Quinlan, Joydeep Ghosh, Qiang Yang, Hiroshi Motoda, Geoffrey J. McLachlan, Angus F. M. Ng, Bing Liu, Philip S. Yu, Zhi-Hua Zhou, Michael Steinbach, David J. Hand, and Dan Steinberg. Top 10 algorithms in data mining. *Knowledge and Information Systems*, 14(1):1–37, 2008.
128. Andrey Gubichev and Thomas Neumann. Exploiting the query structure for efficient join ordering in SPARQL queries. In *Proc. of the 17th International Conference on Extending Database Technology (EDBT 2014)*, pages 439–450, Athens, Greece, March 2014.
129. Chih-Chung Chang and Chih-Jen Lin. LIBSVM: A library for support vector machines. *ACM Transactions on Intelligent Systems and Technology*, 2(3):27, 2011.
130. Daniel D Lee and H Sebastian Seung. Learning the parts of objects by non-negative matrix factorization. *Nature*, 401(6755):788–791, 1999.
131. Thomas Neumann and Guido Moerkotte. Characteristic sets: Accurate cardinality estimation for RDF queries with multiple joins. In *Proc. of the 27th International Conference on Data Engineering (ICDE 2011)*, pages 984–994, Hannover, Germany, April 2011.
132. Bastian Quilitz and Ulf Leser. Querying distributed rdf data sources with sparql. In *Proc. of the 5th Extended Semantic Web Conference (ESWC 2008)*, pages 524–538, Tenerife, Spain, June 2008.
133. Wei Emma Zhang, Mingkui Tan, Quan Z. Sheng, and Qingfeng Shi. Efficient Orthogonal Non-negative Matrix Factorization over Stiefel Manifold. In *Proc. of the 25th ACM International Conference on Information and Knowledge Management (CIKM 2016)*, pages 1743–1752, Indianapolis, IN, USA, October 2016.
134. Pentti Paatero and Unto Tapper. Positive Matrix Factorization: A Non-negative Factor Model with Optimal Utilization of Error Estimates of Data Values. *Environmetrics*, 5(2):111–126, 1994.
135. Filippo Pompili, Nicolas Gillis, Pierre-Antoine Absil, and François Glineur. Two algorithms for orthogonal nonnegative matrix factorization with application to clustering. *Neurocomputing*, 141:15–25, 2014.
136. Hongchang Gao, Feiping Nie, Tom Weidong Cai, and Heng Huang. Robust Capped Norm Nonnegative Matrix Factorization: Capped Norm NMF. In *Proc. of the 24th ACM International on Conference on Information and Knowledge Management (CIKM 2015)*, pages 871–880, Melbourne, Australia, October 2015.
137. Yehuda Koren, Robert M. Bell, and Chris Volinsky. Matrix Factorization Techniques for Recommender Systems. *IEEE Computer*, 42(8):30–37, 2009.
138. Chris H. Q. Ding, Tao Li, Wei Peng, and Haesun Park. Orthogonal Nonnegative Matrix Tri-factorizations for Clustering. In *Proc. of the 12th ACM SIGKDD International Conference on Knowledge Discovery and Data Mining (KDD 2006)*, pages 126–135, Philadelphia, USA, August 2006.
139. Chris H. Q. Ding and Xiaofeng He. On the Equivalence of Nonnegative Matrix Factorization and Spectral Clustering. In *Proc. of the 2005 SIAM International Conference on Data Mining (SDM 2005)*, pages 606–610, Newport Beach, USA, April 2005.
140. Yu-Xiong Wang and Yu-Jin Zhang. Nonnegative Matrix Factorization: A Comprehensive Review. *IEEE Transactions on Knowledge and Data Engineering*, 25(6):1336–1353, 2013.
141. Zhao Li, Xindong Wu, and Hong Peng. Nonnegative Matrix Factorization on Orthogonal Subspace. *Pattern Recognition Letters*, 31(9):905–911, 2010.
142. Zhirong Yang and Erkki Oja. Linear and Nonlinear Projective Nonnegative Matrix Factorization. *IEEE Transactions on Neural Networks*, 21(5):734–749, 2010.
143. Seungjin Choi. Algorithms for orthogonal nonnegative matrix factorization. In *Proc. of the International Joint Conference on Neural Networks (IJCNN 2008)*, pages 1828–1832, Hong Kong, China, June 2008.

144. Megasthenis Asteris, Dimitris Papailiopoulos, and Alexandros G. Dimakis. Orthogonal NMF through Subspace Exploration. In *Proc. of the 29th Annual Conference on Neural Information Processing Systems (NIPS 2015)*, pages 343–351, Montreal, Canada, December 2014.

145. Chris H. Q. Ding, Tao Li, and Michael I. Jordan. Convex and Semi-Nonnegative Matrix Factorizations. *IEEE Transactions on Pattern Analysis and Machine Intelligence*, 32(1):45–55, 2010.

146. Pierre-Antoine Absil, Robert E. Mahony, and Rodolphe Sepulchre. *Optimization Algorithms on Matrix Manifolds*. Princeton University Press, 2008.

147. Magnus Rudolph Hestenes and Eduard Stiefel. Methods of Conjugate Gradients for Solving Linear Systems. *Journal of the Research of the National Bureau of Standards*, 49(6):409–436, 1952.

148. William W Hager and Hongchao Zhang. A Survey of Nonlinear Conjugate Gradient Methods. *Pacific Journal of Optimization*, 2(1):35–58, 2006.

149. Bart Vandereycken. Low-rank matrix completion by riemannian optimization. *SIAM Journal on Optimization*, 23(2):1214–1236, 2013.

150. Jorge Nocedal and J Wright Stephen. *Numerical Optimization*. Springer Series in Operations Research and Financial Engineering, Springer, 2006.

151. Jonathan Barzilai and Jonathan M Borwein. Two-Point Step Size Gradient Methods. *IMA Journal of Numerical Analysis*, 8(1):141–148, 1988.

152. Donald Goldfarb, Zaiwen Wen, and Wotao Yin. A Curvilinear Search Method for p-Harmonic Flows on Spheres. *SIAM Journal on Imaging Sciences*, 2(1):84–109, 2009.

153. Hongchao Zhang and William W. Hager. A Nonmonotone Line Search Technique and Its Application to Unconstrained Optimization. *SIAM Journal on Optimization*, 14(4):1043–1056, 2004.

154. Cho-Jui Hsieh and Inderjit S. Dhillon. Fast coordinate descent methods with variable selection for non-negative matrix factorization. In *Proc. of the 17th ACM SIGKDD International Conference on Knowledge Discovery and Data Mining (KDD 2011)*, pages 1064–1072, San Diego, USA, August 2011.

155. Sameer A Nene, Shree K Nayar, Hiroshi Murase, et al. Columbia object image library (COIL-20). Technical report, Columbia University, 1996.

156. Terence Sim, Simon Baker, and Maan Bsat. The CMU Pose, Illumination, and Expression Database. *IEEE Transactions on Pattern Analysis and Machine Intelligence*, 25(12):1615–1618, 2003.

157. Shi Zhong and Joydeep Ghosh. Generative Model-based Document Clustering: A Comparative Study. *Knowledge and Information Systems*, 8(3):374–384, 2005.

158. Martijn van Breukelen, Robert P, W, Duin, David M. J. Tax, and J. E. den Hartog. Handwritten digit recognition by combined classifiers. *Kybernetika*, 34(4):381–386, 1998.

159. Wei Xu, Xin Liu, and Yihong Gong. Document Clustering Based on Non-Negative Matrix Factorization. In *Proc. of the 26th Annual International ACM SIGIR Conference on Research and Development in Information Retrieval (SIGIR 2003)*, pages 267–273, Toronto, Canada, July 2003.

160. Deguang Kong, Chris H. Q. Ding, and Heng Huang. Robust Nonnegative Matrix Factorization using L21-Norm. In *Proc. of the 20th ACM Conference on Information and Knowledge Management (CIKM 2011)*, pages 673–682, Glasgow, United Kingdom, October 2011.

161. L. Lovász and M.D. Plummer. *Matching Theory*. North Holland, Budapest, 1986.

162. Deng Cai, Xiaofei He, and Jiawei Han. Document Clustering Using Locality Preserving Indexing. *IEEE Transactions on Knowledge and Data Engineering*, 17(12):1624–1637, 2005.

163. Reeves Fletcher and Colin M Reeves. Function Minimization by Conjugate Gradients. *The Computer Journal*, 7(2):149–154, 1964.

164. Bin Shen and Luo Si. Non-Negative Matrix Factorization Clustering on Multiple Manifolds. In *Proc. of the 24th AAAI Conference on Artificial Intelligence (AAAI 2010)*, Atlanta, USA, July 2010.

165. Deng Cai, Xiaofei He, Jiawei Han, and Thomas S. Huang. Graph Regularized Nonnegative Matrix Factorization for Data Representation. *IEEE Transactions on Pattern Analysis and Machine Intelligence*, 33(8):1548–1560, 2011.

166. Naiyang Guan, Dacheng Tao, Zhigang Luo, and Bo Yuan. Manifold Regularized Discriminative Nonnegative Matrix Factorization With Fast Gradient Descent. *IEEE Transactions on Image Processing*, 20(7):2030–2048, 2011.

167. Jin Huang, Feiping Nie, Heng Huang, and Chris H. Q. Ding. Robust Manifold Nonnegative Matrix Factorization. *ACM Transactions on Knowledge Discovery from Data*, 8(3):11, 2013.

168. Fuming Sun, Meixiang Xu, Xuekao Hu, and Xiaojun Jiang. Graph Regularized and Sparse Nonnegative Matrix Factorization with Hard Constraints for Data Representation. *Neurocomputing*, 173:233–244, 2016.

169. Wei Emma Zhang, Quan Z. Sheng, Ermyas Abebe, Muhammad Ali Babar, and Andi Zhou. Mining Source Code Topics Through Topic Model and Words Embedding. In *Proc. of 12th International Conference on Advanced Data Mining and Applications (ADMA 2016)*, pages 664–676, Gold Coast, QLD, Australia, December 2016.

170. Stefan Haefliger, Georg Von Krogh, and Sebastian Spaeth. Code Reuse in Open Source Software. *Management Science*, 54(1):180–193, 2008.

171. Girish Maskeri Rama, Santonu Sarkar, and Kenneth Heafield. Mining Business Topics in Source Code using Latent Dirichlet Allocation. In *Proc. of the 1st Annual India Software Engineering Conference (ISEC 2008)*, pages 113–120, Hyderabad, India, February 2008.

172. Sonia Haiduc, Jairo Aponte, Laura Moreno, and Andrian Marcus. On the Use of Automated Text Summarization Techniques for Summarizing Source Code. In *Proc. of the 17th Working Conference on Reverse Engineering (WCRE 2010)*, pages 35–44, Beverly, MA, USA, October 2010.

173. Sonia Haiduc, Jairo Aponte, and Andrian Marcus. Supporting Program Comprehension with Source Code Summarization. In *Proc. of the 32nd ACM/IEEE International Conference on Software Engineering (ICSE 2010)*, pages 223–226, Cape Town, South Africa, May 2010.

174. Giriprasad Sridhara, Lori L. Pollock, and K. Vijay-Shanker. Automatically Detecting and Describing High Level Actions within Methods. In *Proc. of the 33rd International Conference on Software Engineering (ICSE 2011)*, pages 101–110, Waikiki, Honolulu, HI, USA, May 2011.

175. Laura Moreno, Jairo Aponte, Giriprasad Sridhara, Andrian Marcus, Lori L. Pollock, and K. Vijay-Shanker. Automatic Generation of Natural Language Summaries for Java Classes. In *Proc. of the 21st IEEE International Conference on Program Comprehension (ICPC 2013)*, pages 23–32, San Francisco, NC, USA, May 2013.

176. Giriprasad Sridhara, Emily Hill, Divya Muppaneni, Lori L. Pollock, and K. Vijay-Shanker. Towards Automatically Generating Summary Comments for Java Methods. In *Proc. of the 25th IEEE/ACM International Conference on Automated Software Engineering (ASE 2010)*, pages 43–52, Antwerp, Belgium, September 2010.

177. Giriprasad Sridhara, Lori L. Pollock, and K. Vijay-Shanker. Generating Parameter Comments and Integrating with Method Summaries. In *Proc. of the 19th IEEE International Conference on Program Comprehension (ICPC 2011)*, pages 71–80, Kingston, ON, Canada, June 2011.

178. Paige Rodeghero, Collin McMillan, Paul W. McBurney, Nigel Bosch, and Sidney K. D'Mello. Improving Automated Source Code Summarization via An Eye-tracking Study of Programmers. In *Proc. of the 36th International Conference on Software Engineering (ICSE 2014)*, pages 390–401, Hyderabad, India, June 2014.

179. Tomas Mikolov, Kai Chen, Greg Corrado, and Jeffrey Dean. Efficient Estimation of Word Representations in Vector Space. *CoRR*, abs/1301.3781, 2013.

180. Tomas Mikolov, Ilya Sutskever, Kai Chen, Gregory S. Corrado, and Jeffrey Dean. Distributed Representations of Words and Phrases and their Compositionality. In *Proc. of the 27th Annual Conference on Neural Information Processing Systems (NIPS 2013)*, pages 3111–3119, Lake Tahoe, Nevada, United States, December 2013.

181. Jon Louis Bentley. Multidimensional Binary Search Trees Used for Associative Searching. *Communications of the ACM*, 18(9):509–517, 1975.

182. David M. Blei, Andrew Y. Ng, and Michael I. Jordan. Latent Dirichlet Allocation. *Journal of Machine Learning Research*, 3:993–1022, 2003.

183. Michael Röder, Andreas Both, and Alexander Hinneburg. Exploring the Space of Topic Coherence Measures. In *Proc. of the Eighth ACM International Conference on Web Search and Data Mining (WSDM 2015)*, pages 399–408, Shanghai, China, February 2015.

184. Isabelle Guyon and André Elisseeff. An Introduction to Variable and Feature Selection. *Journal of Machine Learning Research*, 3:1157–1182, 2003.

185. Miltiadis Allamanis and Charles A. Sutton. Mining Source Code Repositories at Massive Scale using Language Modeling. In *Proc. of the 10th Working Conference on Mining Software Repositories (MSR 2013)*, pages 207–216, San Francisco, CA, USA, May 2013.

186. Liqiang Niu, Xinyu Dai, Jianbing Zhang, and Jiajun Chen. Topic2Vec: Learning Distributed Representations of Topics. In *Proc. of the International Conference on Asian Language Processing 2015 (IALP 2015)*, pages 193–196, Suzhou, China, October 2015.

187. Ronald Fagin, Ravi Kumar, and D. Sivakumar. Comparing Top K Lists. In *Proc. of the Fourteenth Annual ACM-SIAM Symposium on Discrete Algorithms (SODA 2003)*, pages 28–36, Baltimore, Maryland, USA, January 2003.

188. Hazeline U. Asuncion, Arthur U. Asuncion, and Richard N. Taylor. Software Traceability with Topic Modeling. In *Proc. of the 32nd ACM/IEEE International Conference on Software Engineering (ICSE 2010)*, pages 95–104, Cape Town, South Africa, May 2010.

189. Stacy K. Lukins, Nicholas A. Kraft, and Letha H. Etzkorn. Bug Localization using Latent Dirichlet Allocation. *Information and Software Technology*, 52(9):972–990, 2010.

190. Wei Emma Zhang, Quan Z. Sheng, Jey Han Lau, and Ermyas Abebe. Detecting Duplicate Posts in Programming QA Communities via Latent Semantics and Association Rules. In *Proc. of 26th International World Wide Web Conference (WWW 2017)*, pages 1221–1229, Perth, WA, Australia, April 2017.

191. Kai Wang, Zhaoyan Ming, and Tat-Seng Chua. A Syntactic Tree Matching Approach to Finding Similar Questions in Community-based QA Services. In *Proc. of the 32nd Annual International ACM SIGIR Conference on Research and Development in Information Retrieval (SIGIR 2009)*, pages 187–194, Boston, MA, USA, July 2009.

192. Guangyou Zhou, Yang Liu, Fang Liu, Daojian Zeng, and Jun Zhao. Improving Question Retrieval in Community Question Answering Using World Knowledge. In *Proc. of the 23rd International Joint Conference on Artificial Intelligence (IJCAI 2013)*, pages 2239–2245, Beijing, China, August 2013.

193. Xin Cao, Gao Cong, Bin Cui, and Christian S. Jensen. A Generalized Framework of Exploring Category Information for Question Retrieval in Community Question Answer Archives. In *Proc. of the 19th International World Wide Web Conference (WWW 2010)*, pages 201–210, Raleigh, North Carolina, April 2010.

194. Xin Cao, Gao Cong, Bin Cui, Christian S. Jensen, and Quan Yuan. Approaches to Exploring Category Information for Question Retrieval in Community Question-Answer Archives. *ACM Transactions on Information Systems*, 30(2):7, 2012.

195. Anna Shtok, Gideon Dror, Yoelle Maarek, and Idan Szpektor. Learning from the Past: Answering New Questions with Past Answers. In *Proc. of the 21st World Wide Web Conference (WWW 2012)*, pages 759–768, Lyon, France, April 2012.

196. Lichun Yang, Shenghua Bao, Qingliang Lin, Xian Wu, Dingyi Han, Zhong Su, and Yong Yu. Analyzing and Predicting Not-Answered Questions in Community-based Question Answering Services. In *Proc. of the 25th AAAI Conference on Artificial Intelligence (AAAI 2011)*, San Francisco, California, USA, August 2011.

197. Muhammad Ahasanuzzaman, Muhammad Asaduzzaman, Chanchal K. Roy, and Kevin A. Schneider. Mining Duplicate Questions in Stack Overflow. In *Proc. of the 13th International Conference on Mining Software Repositories (MSR 2016)*, pages 402–412, May 2016.

198. Yun Zhang, David Lo, Xin Xia, and Jianling Sun. Multi-Factor Duplicate Question Detection in Stack Overflow. *Journal of Computer Science and Technology*, 30(5):981–997, 2015.

199. Ke Tao, Fabian Abel, Claudia Hauff, Geert-Jan Houben, and Ujwal Gadiraju. Groundhog Day: Near-Duplicate Detection on Twitter. In *Proc. of the 22nd International World Wide Web Conference (WWW 2013)*, pages 1273–1284, Rio de Janeiro,Brazil, May 2013.

200. Quoc V. Le and Tomas Mikolov. Distributed Representations of Sentences and Documents. In *Proc. of the 31th International Conference on Machine Learning (ICML 2014)*, pages 1188–1196, Beijing, China, June 2014.

201. Jonathan Berant and Percy Liang. Semantic Parsing via Paraphrasing. In *Proc. of the 52nd Annual Meeting of the Association for Computational Linguistics (ACL 2014)*, pages 1415–1425, Baltimore, MD, USA, June 2014.

202. Marie-Catherine De Marneffe, Bill MacCartney, and Christopher D Manning. Generating Typed Dependency Parses From Phrase Structure Parses. In *Proc. of the 5th International Conference on Language Resources and Evaluation (LREC 2006)*, volume 6, pages 449–454, 2006.

203. Jey Han Lau and Timothy Baldwin. An Empirical Evaluation of doc2vec with Practical Insights into Document Embedding Generation. In *Proceedings of the 1st Workshop on Representation Learning for NLP (RepL4NLP 2016)*, pages 78–86, Berlin, Germany, 2016.

204. Chenliang Li, Haoran Wang, Zhiqian Zhang, Aixin Sun, and Zongyang Ma. Topic Modeling for Short Texts with Auxiliary Word Embeddings. In *Proc. of the 39th International ACM SIGIR conference on Research and Development in Information Retrieval (SIGIR 2016)*, pages 165–174, Pisa, Italy, July 2016.

205. Franz Josef Och and Hermann Ney. A Systematic Comparison of Various Statistical Alignment Models. *Computational Linguistics*, 29(1):19–51, 2003.

206. Philipp Koehn, Franz Josef Och, and Daniel Marcu. Statistical phrase-based translation. In *Proc. of the Human Language Technology Conference of the North American Chapter of the Association for Computational Linguistics (NAACL 2003)*, Edmonton, Canada, May 2003.

207. Franz Josef Och and Hermann Ney. The Alignment Template Approach to Statistical Machine Translation. *Computational Linguistics*, 30(4):417–449, 2004.

208. C Fellbaum. WordNet: An Electronic Lexical Database. *MIT Press*, 1998.

209. Michael Collins. Discriminative training methods for hidden markov models: Theory and experiments with perception algorithms. In *Proc. of the Conference on Empirical Methods in Natural Language Processing (EMNLP 2002)*, pages 1–8, Philadelphia, PA, USA, July 2002.

210. Leo Breiman, J. H. Friedman, R. A. Olshen, and C. J. Stone. *Classification and Regression Trees*. Wadsworth, 1984.

211. Marti A. Hearst, Susan T Dumais, Edgar Osman, John Platt, and Bernhard Scholkopf. Support Vector Machines. *IEEE Intelligent Systems and their Applications*, 13(4):18–28, 1998.

212. Strother H Walker and David B Duncan. Estimation of the Probability of an Event as a Function of Several Independent Variables. *Biometrika*, 54(1–2):167–179, 1967.

213. Tin Kam Ho. The Random Subspace Method for Constructing Decision Forests. *IEEE Transactions on Pattern Analysis and Machine Intelligence*, 20(8):832–844, 1998.

214. Tony F Chan, Gene Howard Golub, and Randall J LeVeque. Updating Formulae and A Pairwise Algorithm for Computing Sample Variances. In *Proc. of the 5th Symposium in Computational Statistics (COMPSTAT 1982)*, pages 30–41, Toulouse, France, 1982.

215. Tong Zhang. Solving Large Scale Linear Prediction Problems Using Stochastic Gradient Descent Algorithms. In *Proc. of the 21st International Conference on Machine Learning (ICML 2004)*, pages 919–926, Banff, Alberta, Canada, July 2004.

216. Koby Crammer, Ofer Dekel, Joseph Keshet, Shai Shalev-Shwartz, and Yoram Singer. Online Passive-Aggressive Algorithms. *Journal of Machine Learning Research*, 7:551–585, 2006.

217. Rich Caruana and Alexandru Niculescu-Mizil. An empirical comparison of supervised learning algorithms. In *Proc. of the 23rd International Conference on Machine Learning (ICML 2006)*, pages 161–168, Pittsburgh, Pennsylvania, USA, June 2006.

218. Blerina Bazelli, Abram Hindle, and Eleni Stroulia. On the Personality Traits of StackOverflow Users. In *Proc. of the 29th IEEE International Conference on Software Maintenance (ICSM 2013)*, pages 460–463, Eindhoven, The Netherlands, September 2013.

219. Denzil Correa and Ashish Sureka. Chaff from the Wheat: Characterization and Modeling of Deleted Questions on Stack Overflow. In *Proc. of the 23rd International World Wide Web Conference (WWW 2014)*, pages 631–642, Seoul, Republic of Korea, April 2014.

220. Christoph Treude, Ohad Barzilay, and Margaret-Anne D. Storey. How Do Programmers Ask and Answer Questions on the Web? In *Proc. of the 33rd International Conference on Software Engineering (ICSE 2011)*, pages 804–807, Waikiki, Honolulu, HI, USA, May 2011.

221. Wei Emma Zhang, Ermyas Abebe, Quan Z. Sheng, and Kerry L. Taylor. Towards Building Open Knowledge Base From Programming Question-Answering Communities. In *Proc. of the 15th International Semantic Web Conference (ISWC 2016). Posters & Demonstrations Track*, Kobe, Japan, October 2016.

222. Andreas Holzinger, Christof Stocker, Bernhard Ofner, Gottfried Prohaska, Alberto Brabenetz, and Rainer Hofmann-Wellenhof. *Combining HCI, Natural Language Processing, and Knowledge Discovery-Potential of IBM Content Analytics as an Assistive Technology in the Biomedical Field*, pages 13–24. Springer, 2013.

223. Sangha Nam, YoungGyun Hahm, Sejin Nam, and Key-Sun Choi. SRDF: korean open information extraction using singleton property. In *Proc. of the 14th International Semantic Web Conference (ISWC 2015), Posters & Demonstrations Track*, Bethlehem, USA, October 2015.

Printed in the United States
By Bookmasters